Coloring Atlas of HUMAN ANATOMY

SECOND EDITION

Edwin Chin Jr., Ph.D.
San Jose´ State University

Joanne Thorner Kerr, Ph. D.
San Jose´ State University

SAUNDERS COLLEGE PUBLISHING
Harcourt Brace College Publishers
Fort Worth Philadelphia San Diego New York
Orlando Austin San Antonio Toronto
Montreal London Sydney Tokyo

Printed in the United States of America.

Chin/Kerr: Coloring Atlas of Human Anatomy, Second Edition

ISBN 0-03-001806-4

345 021 987654321

Preface

This Coloring Atlas of Human Anatomy has been designed to assist students at all levels in their mastery of Anatomical Principles. Anatomy is truly a visual science that is best appreciated both on a Gross and Microscope level when the individual can indeed *picture* the organization of the tissues. Once the fundamental organization is clear, then considerations of related functions through the study of Physiology becomes increasingly easier. Next in the progression of learning will be the examination of disease states (Pathology) and drug interactions (Pharmacology), all made more understandable with a thorough knowledge of Anatomy.

Students are encouraged to utilize the Coloring Atlas to enhance and reinforce the learning gained from textbooks, lectures and laboratory exercises. Each plate has been carefully drawn to clearly demonstrate the features present, while the accompanying legends have been both expanded and simplified in this second edition. As the student uses colored pencils to complete each illustration, they will find that the principles are reinforced and that there is actually something quite enjoyable about coloring! Once a series of plates have been completed, the student may then utilize the Coloring Atlas to test their knowledge of the material. We recommend colored pencils (rather than pens) be used (so that mistakes may be easily removed) and that the student select representative colors. Traditionally, brown has been used for muscles, green for ligaments, red for arteries, blue for veins and yellow for nerves. The student ought to select colors that they like best and strive to make each plate artistically pleasing. We suggest that when a structure is colored on the illustration, that the corresponding color be used on the legend key (the circled numbers at the bottom of each plate description).

Dr. Chin produced all of the illustrations in this book and great care was extended to ensure that the figures would be both clear and accurate. Dr. Kerr wrote the accompanying legends with the hope that students would find the descriptions helpful in their total understanding of Anatomy. The Coloring Atlas of Human Anatomy is intended to be both educational as well as fun, bringing the student to a greater appreciation of King David's observation that we are 'fearfully and wonderfully made"!

Contents

Contents, continued

Anatomical Position, Terms and Planes of the Body

The **Anatomical Position** refers to a standardized view of the body in which the figure is upright with the arms at the sides, palms facing upward (anterior) and the eyes looking forward. The Anatomical Position is divided into various **Anatomical Planes** of the body, indicated by the arrows and dotted lines. The **Midsagittal (or Median) Plane (1)** is the vertical line drawn down the midline of the body, dividing the body into right and left halves. Any vertical line drawn parallel to this Midsagittal Plane is referred to as a **Parasagittal (or Paramedian) Plane (2)**. The **Coronal (or Frontal) Plane (3)** is a vertical line perpendicular to the Midsagittal Plane which divides the body into front and rear portions. The **Transverse (or Horizontal) Plane (4)** divides the upper from lower portions of the body. Various **Anatomical Terms** are utilized to describe the position of structures within the body. For example, **Anterior (or Ventral) (5)** and **Posterior (or Dorsal) (6)** refer to the structures that lie near the front and back of the body, respectively. **Superior (or Cranial) (7)** refers to a structure that is located nearer to the head, while **Inferior (or Caudal) (8)** indicates a position closer to the feet (or tail!). A structure that is **Medial (9)** lies closer to the Midsagittal Plane, while a more **Lateral (10)** structure is positioned farther away from the midline. With reference to the limbs, a structure located near the trunk is **Proximal (11)** (in close proximity), while a structure situated closer to the end of the limb is considered to be **Distal (12)** (more distant).

(1) Midsagittal Plane
(2) Parasagittal Plane
(3) Coronal Plane
(4) Transverse Plane
(5) Anterior
(6) Posterior
(7) Superior
(8) Inferior
(9) Medial
(10) Lateral
(11) Proximal
(12) Distal

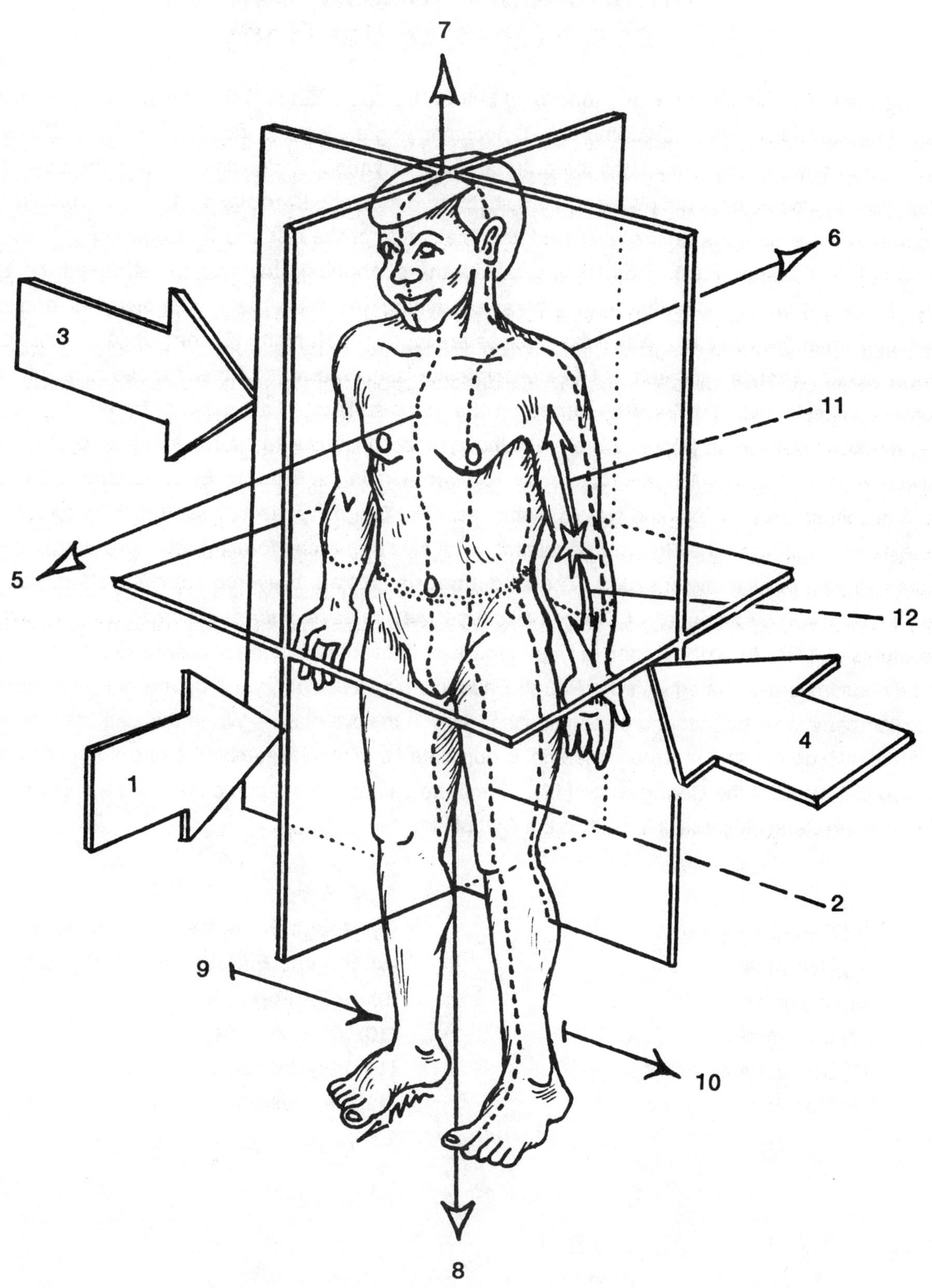

IN-1

Typical Mammalian Cell

The Cell is the basic structural and functional unit of the body. Each Cell consists of four fundamental elements: the Cell (or Plasma) Membrane, Cytoplasm, Organelles and Inclusions. The **Cell Membrane (1)** is a thin envelope that regulates the entry and exit of various molecules found in the cell. The Cell Membrane encloses the liquid phase of the cell, referred to as the **Cytoplasm (2)** which contains various Organelles. The **Nucleus (3)** is often found near the center of the Cell and includes the genetic material found in the **Chromatin (4)**. The limiting membrane surrounding the Nucleus is called the **Nuclear Membrane (5)** which has numerous pores, permitting the movement of molecules between the Nucleoplasm and Cytoplasm. Within the Nucleus are found the **Nucleoli (6)** composed of ribosomal RNA concerned with Protein synthesis. An extensive cytoplasmic network of membranes is also involved in protein synthesis and contains **Ribosomes** on the outer surface. This network is termed the **Rough Endoplasmic Reticulum (7)** and it is often found to be continuous with the Nuclear Membrane. A similar network of membranes which lack Ribosomes is referred to as the **Smooth Endoplasmic Reticulum (8)** and it is concerned with lipid and steroid synthesis. The **Golgi Apparatus (9)** functions to package the proteins released from the Rough Endoplasmic Reticulum, thereby forming the cytoplasmic **Vacuoles**, made ready for cellular use or export. The **Mitochondria (10)** are bilayered spherical structures that are said to be the "powerhouse" of the cell because they release energy which is used to synthesize molecules of ATP. **Polyribosomes (11)** are groups of ribosomes that are similar in structure and function to the ribosomes associated with the Rough Endoplasmic Reticulum, yet they are not associated with a reticular membrane, but rather are seen floating freely in the cytoplasm. When this cell is ready to divide and form two new cells, the genetic material is duplicated and then assembled along a spindle formed by the two Centrioles of the **Centrosome (12)**. Once the cell division is completed, the spindle disappears with only the Centrioles remaining within the Cytoplasm.

(1) Cell Membrane
(2) Cytoplasm
(3) Nucleus
(4) Chromatin
(5) Nuclear Membrane
(6) Nucleoli
(7) Rough Endoplasmic Reticulum
(8) Smooth Endoplasmic Reticulum
(9) Golgi Apparatus
(10) Mitochondria
(11) Polyribosomes
(12) Centrosome

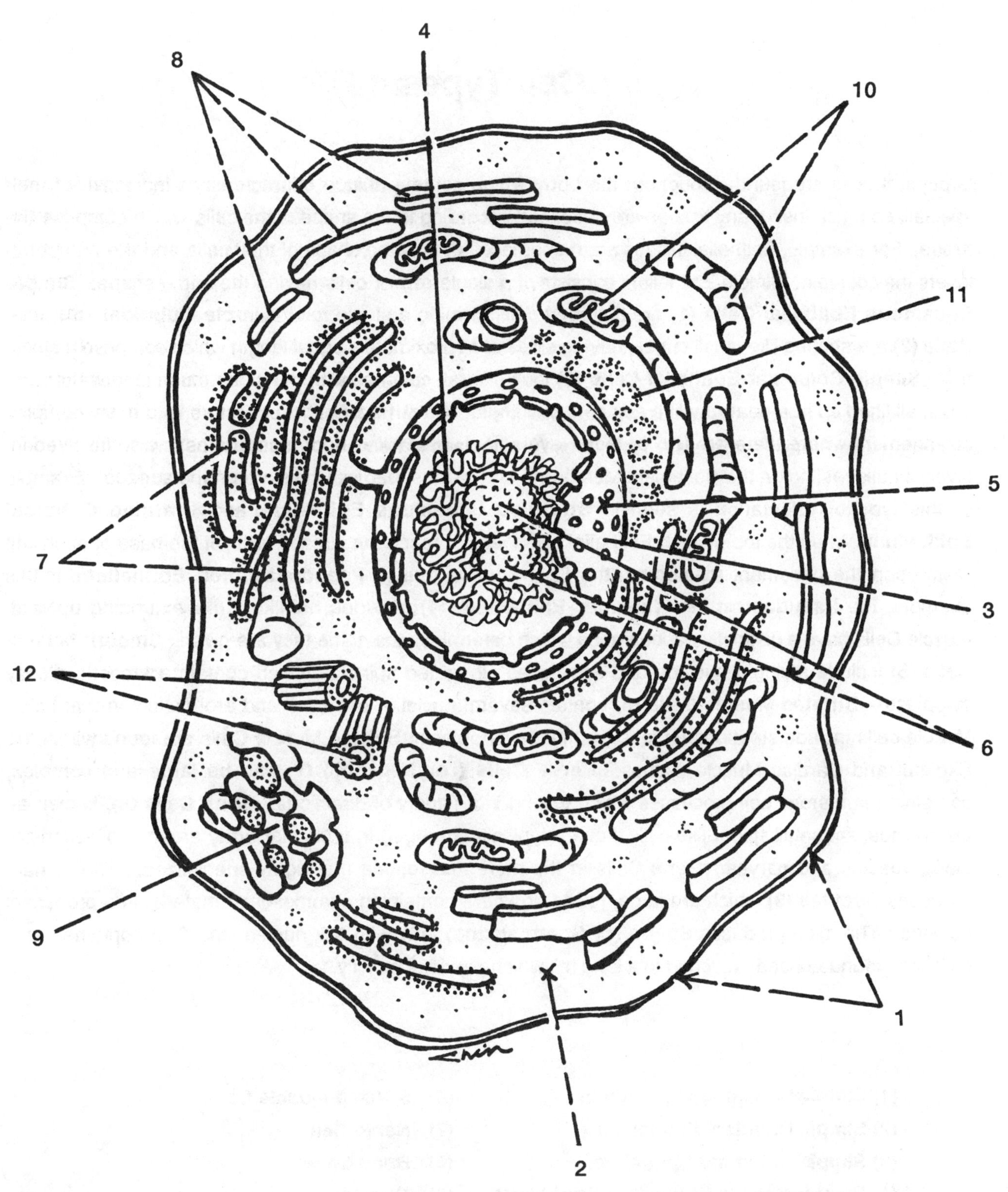
4
8
10
11
5
7
3
12
6
9
1
2

Cell Types

Groups of cells are found throughout the body which contain unique characteristics indicative of their specialized functions. Many tissues are described according to the shape of the cells which comprise the tissue. For example, Epithelial tissues are named according to the shape of their cells and the number of layers they contain. Simple Epithelium consists of a single row of cells having the same shape: **Simple Squamous Epithelial Cells (1)** are flattened cells forming a single row. **Simple Cuboidal Epithelial Cells (2)** are shaped like small cubes, having sides of approximately equal length which comprise a single row. **Simple Columnar Epithelial Cells (3)** look like tall columns with the sides much longer than the base, all lined up in a neat row like adjacent buildings. **Stratified** Epithelium refers to a more complex arrangement where there are two or more layers of cells stacked upon one another, with the deepest layer of cells resting on the Basement Membrane and more superficial cells are on the surface. Example of this type of orientation is seen in **Stratified Squamous Epithelium** and **Stratified Cuboidal Epithelium.** When the top of each Epithelial cell does not reach the surface and yet the base of each cell rests upon the basement membrane, the Epithelium is termed **Pseudostratified Epithelium.** In this example, the **Pseudostratified Columar Epithelium (4)** includes hairlike **Cilia** extending upward. Muscle Cells have a distinctive appearance which determines the name they are given. **Smooth Muscle Cells (5)** include a single nucleus and appear as elongated spindles which contain a smooth, glossy cytoplasm. **Striated Muscle Cells (6)** contain several nuclei in each cell and are distinct in that these Muscle cells include striations (stripes) within the cytoplasm. Striated Muscle Cells are found within the Skeletal and Cardiac Muscle tissues. **Nerve Cells (or Neurons) (7)** can be large and complex, containing numerous cell processes which extend in a variety of directions. **Bone Cells (8)**, known as Osteocytes, are small spider-like cells that are typically arranged in a circular array around collections of blood vessels and nerves. Germ Cells in the male and female are highly specialized. The female produces **Oocytes (9)** which are large, round egg cells containing chromosomal material in a prominent nucleus. The male produces **Sperm (or Spermatozoa) (10)** with the nuclear material contained in a compact, rounded head attached to a long tail which provides motility.

(1) Simple Squamous Epithelium
(2) Simple Cuboidal Epithelium
(3) Simple Columnar Epithelium
(4) Pseudostratified Columnar Epithelium
(5) Smooth Muscle Cells
(6) Striated Muscle Cells
(7) Nerve Cell
(8) Bone Cells
(9) Oocyte
(10) Sperm

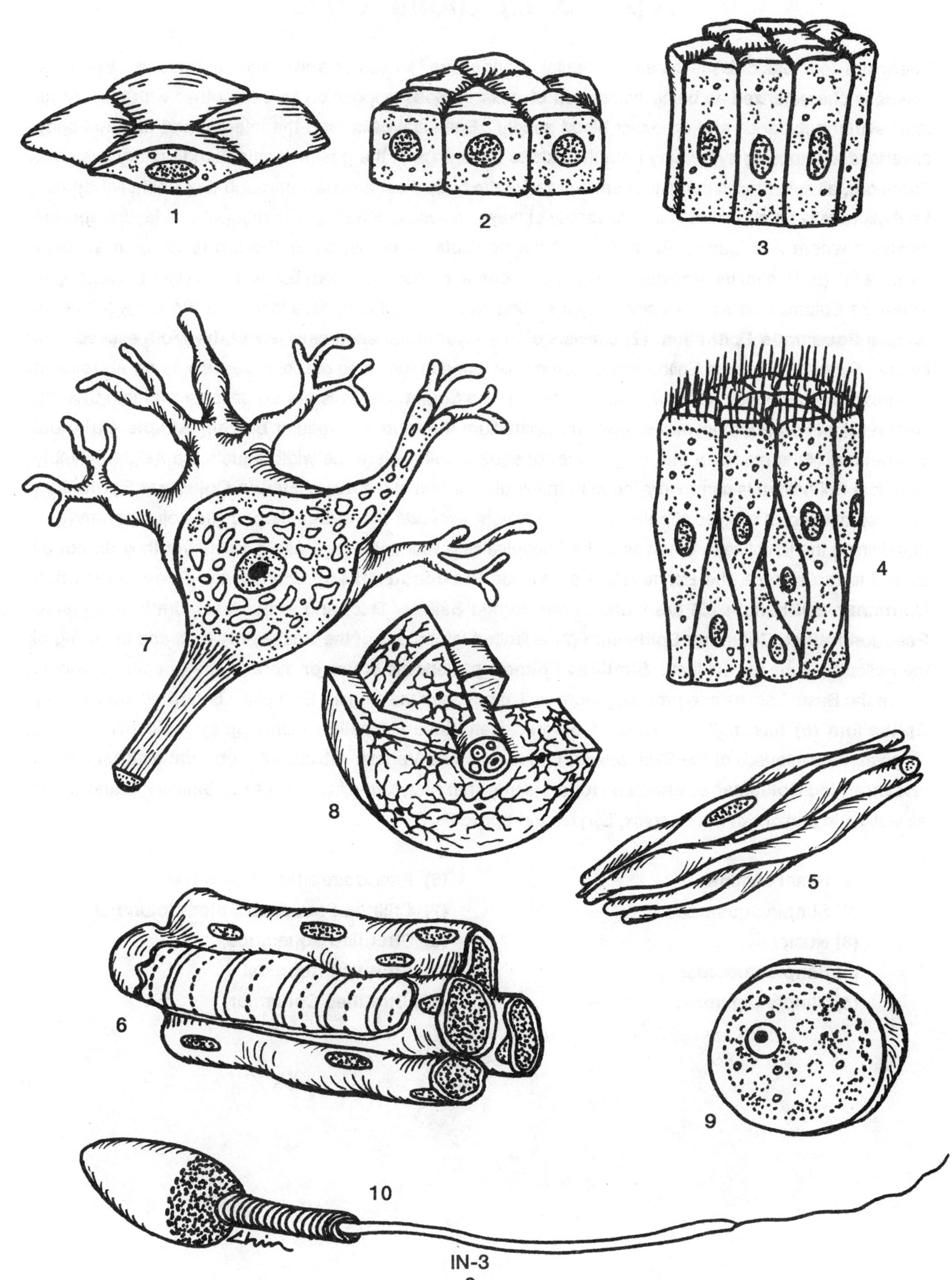

IN-3

Types of Epithelial Cells

There are four basic tissue types: Epithelial, Connective, Muscular and Nervous Tissues. Epithelial Tissue is characterized as being composed of cells in close apposition to each other with little or no intercellular substance and no direct blood supply. Epithelial cells form the internal and external body coverings, comprising everything from the skin to the linings of the gastrointestinal and urinary systems. Therefore, all substances that either enter or leave the body must first pass through this Epithelial barrier. Epithelium may be classified in a wide variety of ways, including cell shape, number of cell layers and the locations where it is found. Regardless of the particular features, all Epithelium rests upon a **Basal Lamina (1)** (or Basement Membrane) which provides a supporting foundation with considerable strength. When the Epithelial cells form a single layer resting on this Basal Lamina, it is termed **Simple** Epithelium. **Simple Squamous Epithelium (2)** consists of one layer of flattened cells where the width exceeds the height. Simple Squamous Epithelium is commonly found in the lining of Blood Vessels, Lung Alveoli and in portions of the Kidney Nephron. Occasionally, these Squamous Cells are so delicate and thin that the **Nuclei (3)** will protrude upward to give the Epithelium the form an irregular border. **Simple Cuboidal Epithelium (4)** is made up of a single layer of square cells where the width equals the height. Simple Cuboidal Epithelium is commonly found in many glands and in the liver. **Simple Columnar Epithelium (5)** is composed of a layer of cells where the height exceeds the width (shaped like columns) and are found lining the Intestines. When all of the Epithelial cells rest upon the Basal Lamina yet they do not all extend to the surface, the Epithelium is said to be **Pseudostratified**. Examples of **Pseudostratified Columnar Epithelium (6)** are found in the Parotid Salivary Duct and in the Male Urethra. **Ciliated Pseudostratified Columnar Epithelium (7)** is limited to portions of the Respiratory tract and the lining of the Fallopian Tube (or Oviduct). **Stratified Epithelium** consists of two or more layers of cells extending above the Basal Lamina and providing more resistance than the Simple Epithelia. **Stratified Squamous Epithelium (8)** has multiple layers of progressively flattened cells which display rapid mitosis and renewal, characteristic of the Skin, Mouth, Esophagus and Vagina. **Stratified Cuboidal Epithelium (9)** and **Stratified Columnar Epithelium (10)** are somewhat rare, found in some of the Salivary Gland Ducts as well as in portions of the Pharynx, Larynx and Epiglottis.

(1) Basal Lamina
(2) Simple Squamous
(3) Nuclei
(4) Simple Cuboidal
(5) Simple Columnar
(6) Pseudostratified Columnar
(7) Ciliated Pseudostratified Columnar
(8) Stratified Squamous
(9) Stratified Cuboidal
(10) Stratified Columnar

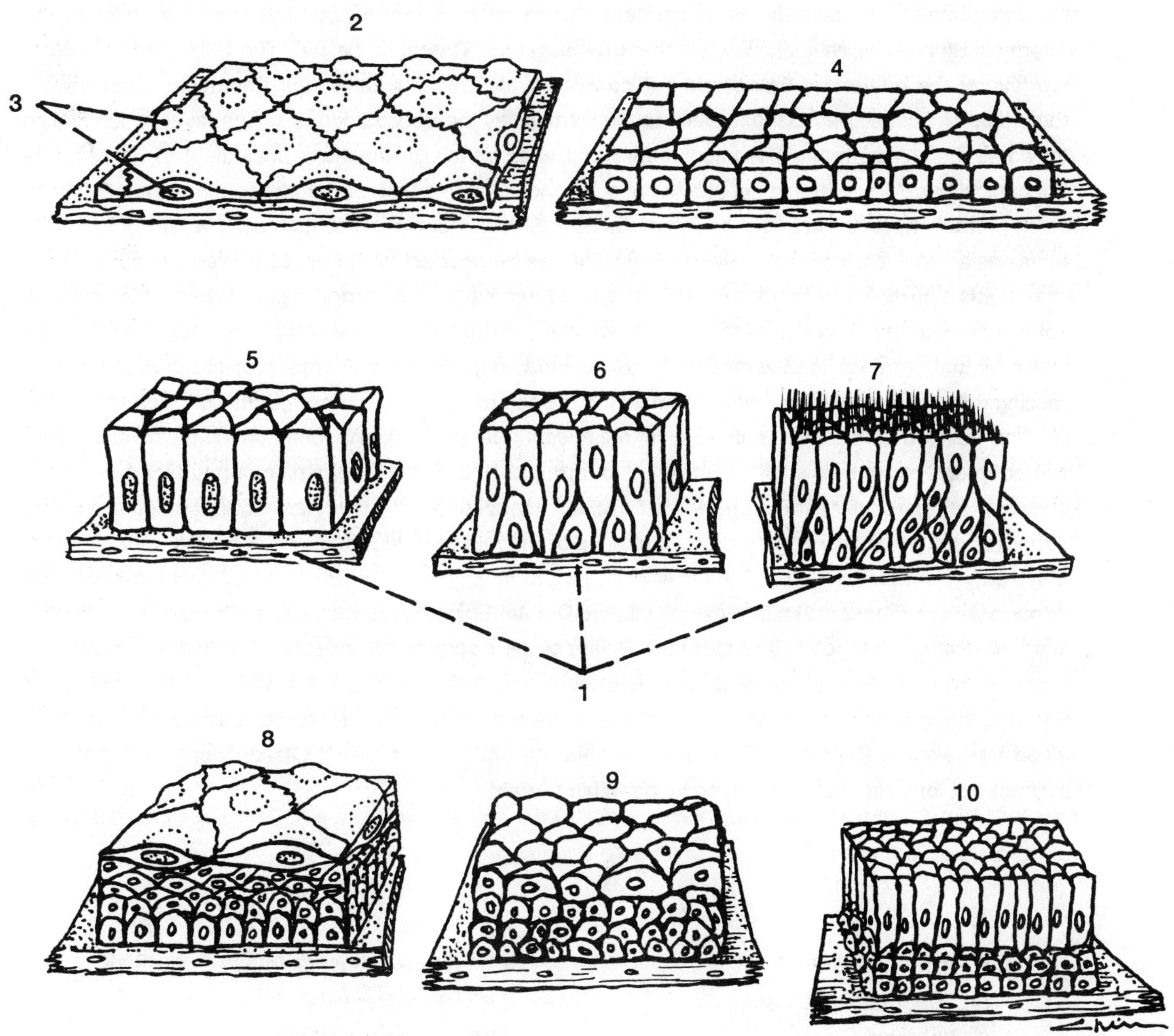
2
3
4
5
6
7
1
8
9
10

Integument

The Integument is a covering or investment that extends over the body surfaces. It includes the outermost **Skin (1)** which is anchored to the **Subcutaneous Tissue (2)** below. The Skin is comprised of two layers, the outermost **Epidermis (3)** consisting of Epithelial Cells, the underlying **Dermis (4)** comprised of Connective Tissues. Deep to the Dermis lies the Subcutaneous Tissue layer which is also known as the Hypodermis. The Integument has a wide variety of functions, including: 1) serves as a mechanical barrier, preventing penetration of substances into the body 2) produces the pigment Melanin which protects against Ultraviolet radiation 3) assists in the excretion of metabolic waste products 3) serves as a shock absorber 4) protects against injuries from impact or friction 5) prevents evaporation of body fluids 6) helps in thermoregulation and 7) serves as a sensory organ. Beginning with the Epidermis, note the presence of the outermost layer or **Stratum Corneum (5)** which consists of tough, protective keratinized cells. **Sweat Glands (6)**, located deep within the Dermis, project a thin Sweat Duct passing through the Epidermis which permits the release of sweat onto the Stratum Corneum. **Hairs (7)** are also included on most skin surfaces, yet are absent from the thick skin found on the palms and soles. The deep end of the Hair shaft is expanded to form the **Hair Follicle (8)** which is nourished by **Blood Vessels (9)** present in the Superficial Fascia. Associated with the Hair Shaft are oil-producing **Sebaceous Glands (10)** and specialized strands of Smooth Muscle termed **Arrector Pili Muscles (11)** which serve to alter the angle of the Hair shaft (to produce "goose bumps"). Notice the presence of **Sensory Nerve Fibers (12)** which extend into the Dermal-Epidermal junction, and are responsive to pain and temperature changes. Note also the location of the **Pacinian Corpuscle (13)** which is sensitive to pressure and vibration. There is abundant insulating **Adipose Tissue (14)** found throughout the Subcutaneous Tissue which serves to insulate, store energy and acts as a shock aborber. Numerous Blood Vessels and Nerves, as well as a few Apocrine and Eccrine Sweat Glands along with the lower segments of long Hair Follicles can be found within Dermis.

(1) Skin
(2) Subcutaneous Tissue
(3) Epidermis
(4) Dermis
(5) Stratum Corneum
(6) Sweat Gland
(7) Hairs

(8) Hair Follicle
(9) Blood Vessels
(10) Sebaceous Gland
(11) Arrector Pili Muscle
(12) Sensory Nerve Fibers
(13) Pacinian Corpuscle
(14) Adipose Tissue

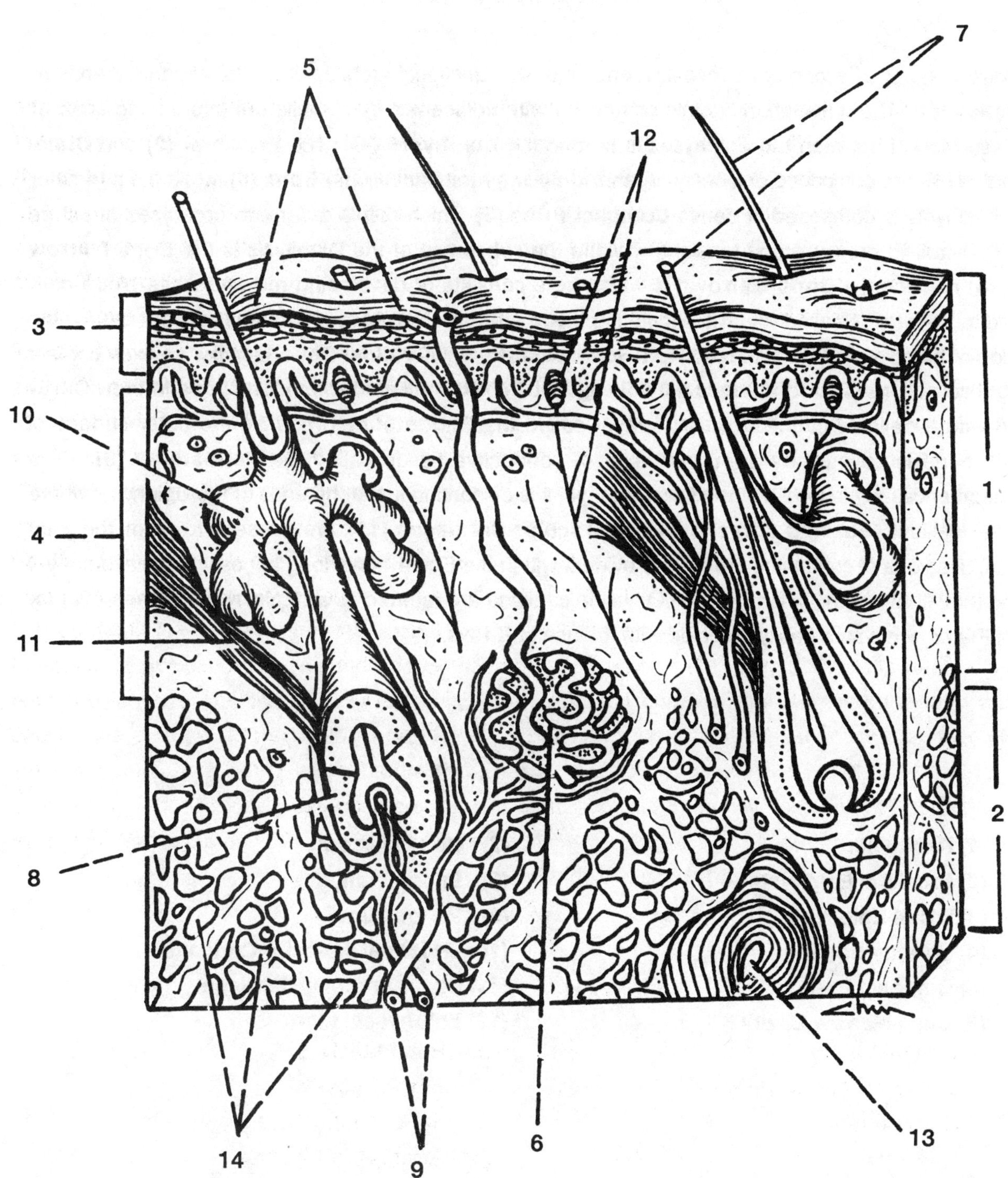
7
5
12
3
10
4
11
8
1
2
14
9
6
13
Chin

Organization of Bone

The gross anatomy of a long bone reveals a characteristic functional architecture. The expanded ends or **Epiphyses** form the articulations (joints or unions) with adjacent bones, while the intervening shaft of bony tissue found between the Epiphyses is termed the **Diaphysis (1)**. The **Proximal (2)** and **Distal Epiphyses (3)** are composed of loosely organized **Spongy** (or Cancellous) **Bone (4)**, while the perimeter of the Diaphysis is composed of dense **Compact Bone (5)** which displays a more organized structure that is resistant to pressure and tensions. Within the substance of the Diaphysis is the **Bone Marrow Cavity (6)** housing the **Bone Marrow (7)** which here consists of developing red and white blood cells destined to be later released into the circulating blood stream. Bone Marrow that is responsible for the formation of blood elements is termed **Red Bone Marrow**, while **Yellow Bone Marrow** cavities contain abundant collections of fat (and hence are not engaged in blood cell production). The Bone Marrow Cavity is lined with a delicate connective tissue layer called the **Endosteum (8)**, while the outer surface of compact bone is covered by a tougher layer of connective tissue called the **Periosteum (9)**. The Periosteum encases the length of the Diaphysis, yet it is discontinuous at the ends of the Epiphysis where the more resilient **Proximal (10)** and **Distal Articular Cartilages (11)** are present to form the joint surfaces. Regions of continuing growth are evident within the Epiphysis of long bones which have not yet reached their final length. These zones of Hyaline cartilage are referred to as **Epiphyseal Lines (12)** (or Growth Plates) which are easily detected with the use of X-rays.

(1) Diaphysis
(2) Proximal Epiphysis
(3) Distal Epiphysis
(4) Spongy Bone
(5) Compact Bone
(6) Bone Marrow Cavity
(7) Bone Marrow
(8) Endosteum
(9) Periosteum
(10) Proximal Articular Cartilage
(11) Distal Articular Cartilage
(12) Epiphyseal Line

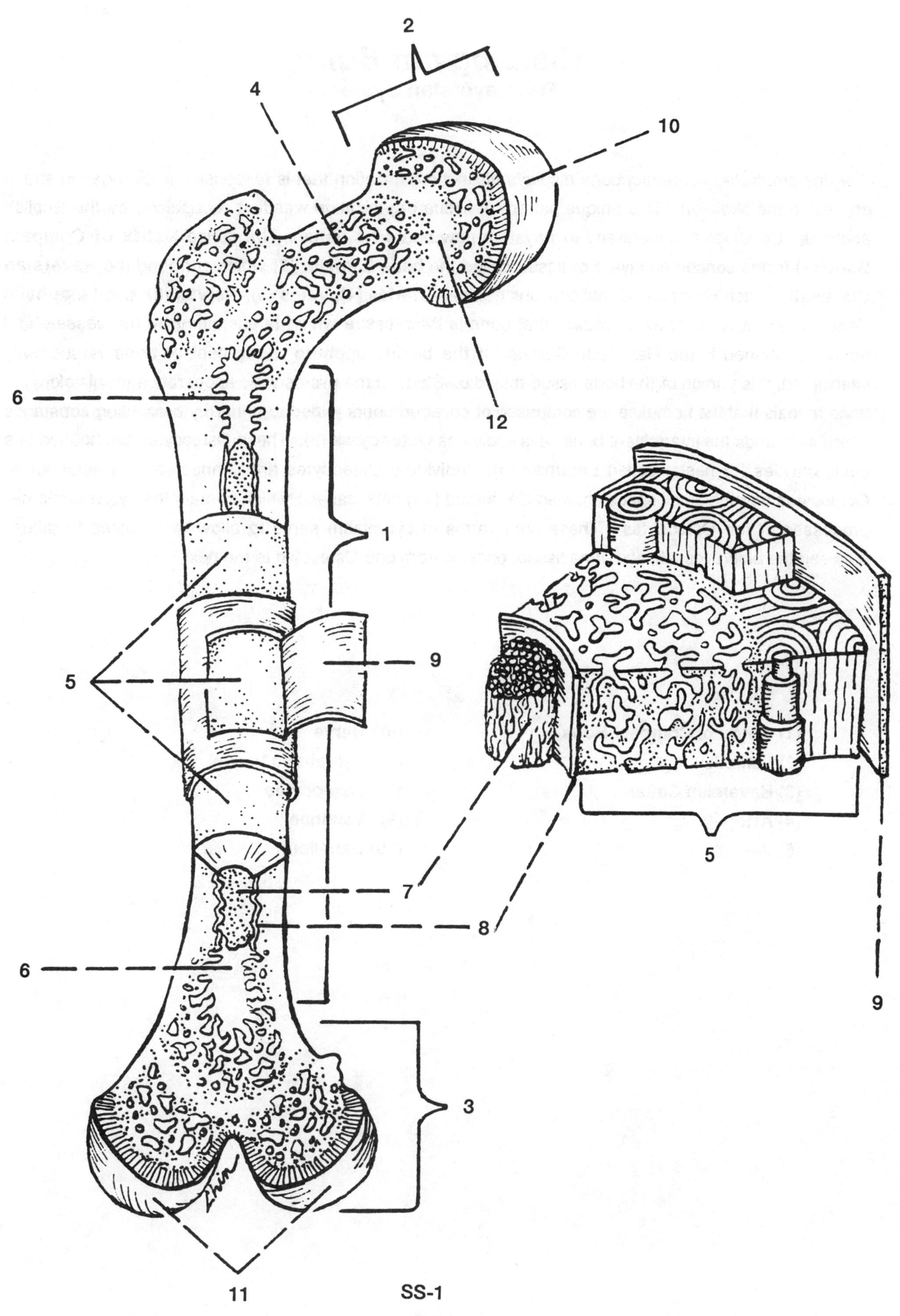

SS-1

Histology of Bone
The Haversian System

The fine structure of compact bone displays distinct organization that is responsive to changes in stress applied to the skeleton. This unique pattern of tissue organization was first recognized by the English anatomist Dr. Clopton Havers and this system now bears his name. The dense **Matrix of Compact Bone (1)** forms concentric layers of tissue called the bone **Lamella (2)** which surround the **Haversian Canals (3).** Each Haversian Canal contains nutrient **Arteries (4), Veins (5), Nerves (6),** and **Lymphatic Vessels (7).** It is important to realize that bone is living tissue which is sustained by the vessels and nerves contained in the Haversian Canals. If the blood supply to a segment of bone is suddenly interrupted, this portion of the bone tissue may die. Study of the microscopic appearance (or histology) of bone reveals that the Lamellae are comprised of collagen fibers joined together by cementing substance which surrounds the intermittent bone cells known as **Osteocytes (8).** These Osteocytes are housed in a small cavities (or nests) called **Lacunae (9).** Individual Osteocytes are connected with neighboring Osteocytes through small spaces named **Canaliculi (10)** (little canals) which contain the cytoplasmic cell processes of the Osteocytes. These small arms of cytoplasm serve to allow metabolites to diffuse between the blood vessels and bone tissue, passing from one Osteocyte to the next.

(1) Matrix of Compact Bone
(2) Lamella
(3) Haversian Canal
(4) Artery
(5) Vein
(6) Nerve
(7) Lymphatic Vessel
(8) Osteocytes
(9) Lacunae
(10) Canaliculi

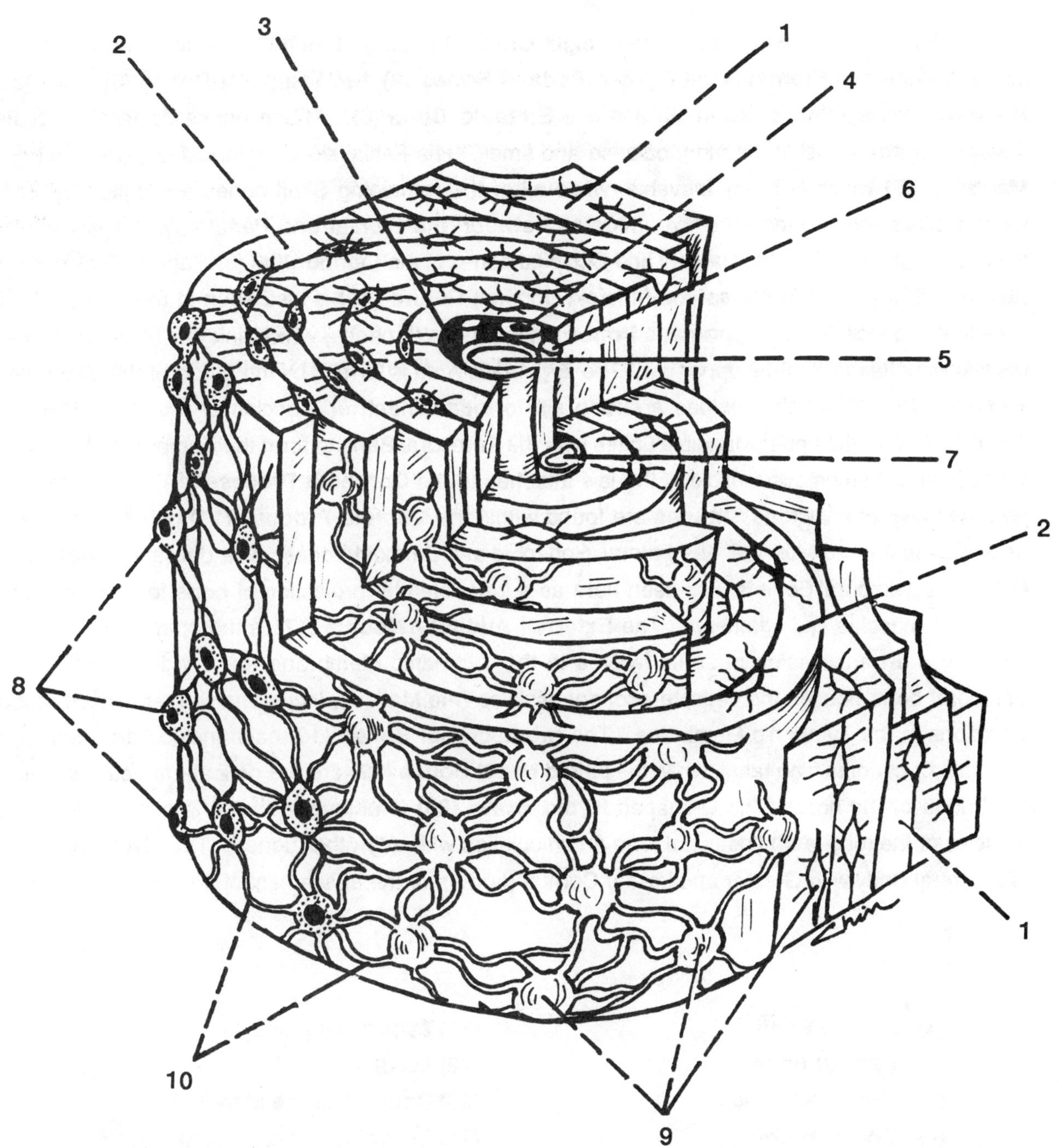

1
2
3
4
5
6
7
8
9
10

Human Skull
(Lateral View)

The Skull consists of twenty two bones, eight Cranial bones and fourteen Facial bones. The Cranial bones include one **Frontal Bone (1)**, two **Parietal Bones (2)**, two **Temporal Bones (3)**, one **Occipital Bone (4)**, one **Sphenoid Bone (5)** and one **Ethmoid Bone (6)**. These bones protect the brain and sensory organs of vision, hearing, balance and smell. The Facial skeleton includes lower jaw bone, the **Mandible (7)** which is freely movable, while all of the remaining Skull bones are united by **Sutures**, jagged joints with limited mobility. Situated between the Frontal and Parietal bones is the **Coronal Suture (8)** and between the Parietal and Occipital bones is the **Lambdoidal** (or Parieto-Occipital) **Suture (9)**. In addition, the **Squamosal** (or Parieto-Temporal) **Suture (10)** is also seen in this view. The Facial skeleton also includes the **Zygomatic Bones (11)** (the cheek bones) which link the Temporal Bones with the two **Maxillae (12)** of the upper jaw. The Zygomatic Arch is formed by the union of the Zygomatic and Temporal Bones, which provides a strong site of attachment for muscles of the jaw. The rounded **Condyle (13)** of the upper Mandible unites with the Temporal Bone to form the Temporomandibular Joint. The powerful Temporalis muscle of the jaw attaches to the **Coronoid Process (14)**, acting to close the jaw. Air Cavities known as **Sinuses** are found within the **Mastoid Process (15)** of the Temporal Bone, as well as in the two Maxillae, the Frontal, Sphenoid and Ethmoid bones. The delicate **Styloid Process (16)** of the Temporal Bone is also seen here as a long, slender projection of bone for the attachment of important muscles and ligaments. The **External Auditory Meatus (17)** is the opening in the Temporal bone for the transmission of sound waves to the Tympanic Membrane (or "Eardrum"). The resilient Temporal bone houses the delicate auditory ossicles (the Malleus, Incus and Stapes) which respond to vibrations in the Tympanic Membrane. The two **Lacrimal Bones (18)** each include an opening for the Lacrimal Duct containing tears, while the paired **Nasal Bones (19)** are the delicate structures which form the bridge of the nose. The U-shaped **Hyoid Bone (20)** is unique in that it is suspended within the muscles of the tongue and does not form an articulation with any other bones. The Hyoid Bone consists of a central Body with Greater and Lesser Cornua (horns) for the attachment of the tongue musculature.

(1) Frontal Bone
(2) Parietal Bone
(3) Temporal Bone
(4) Occipital Bone
(5) Sphenoid Bone
(6) Ethmoid Bone
(7) Mandible
(8) Coronal Suture
(9) Lambdoidal Suture
(10) Squamosal Suture
(11) Zygomatic Bone
(12) Maxilla
(13) Condyle of the Mandible
(14) Coronoid Process of Mandible
(15) Mastoid Process of Temporal Bone
(16) Styloid Process of Temporal Bone
(17) External Auditory Meatus of Temporal Bone
(18) Lacrimal Bone
(19) Nasal Bone
(20) Hyoid Bone

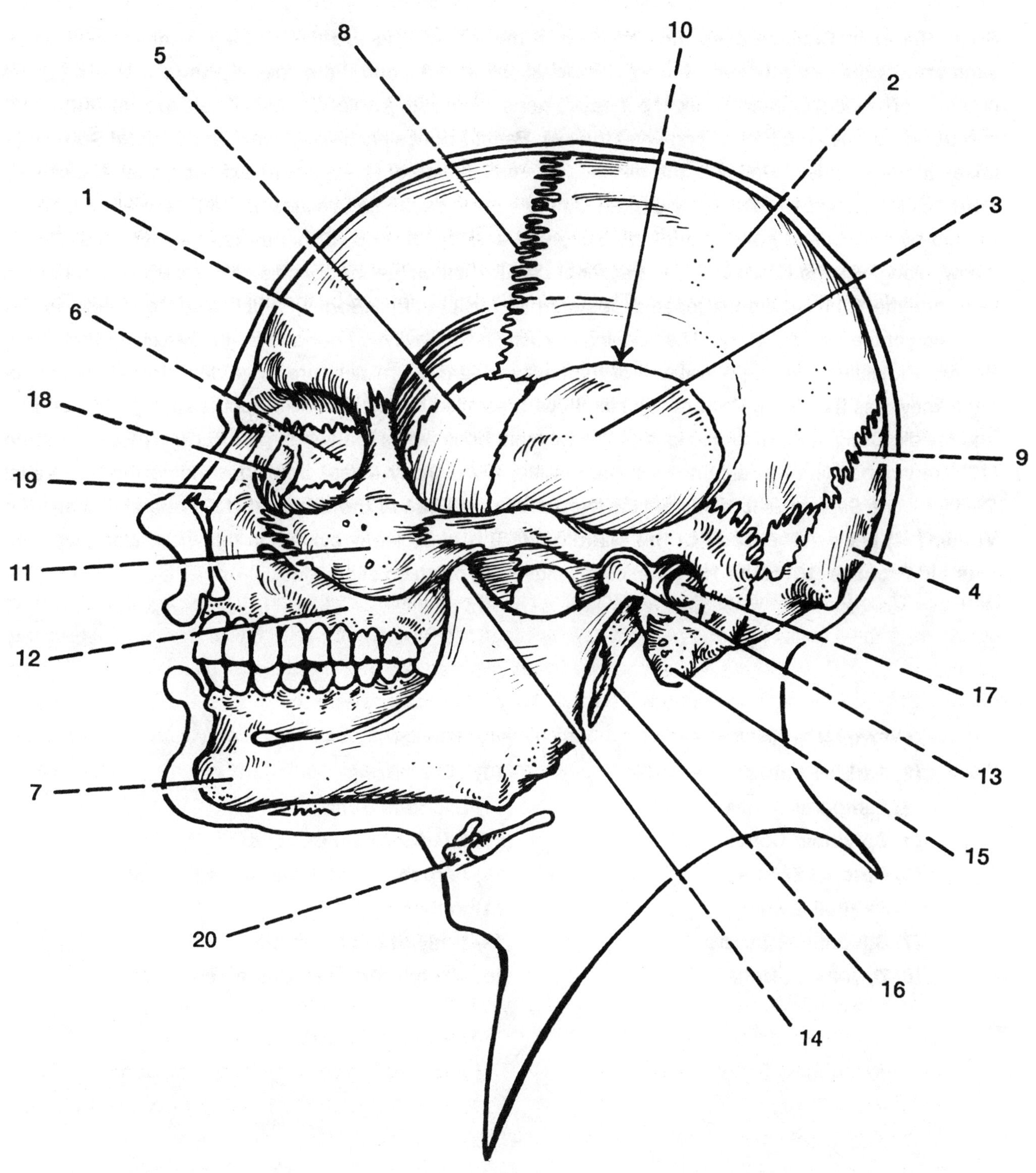

SS-3

Human Skull

(Anterior View)

Six of the eight Cranial bones may be seen in this view: The **Frontal Bone (1)**, along with small segments of the two **Parietal (2)** and **Temporal Bones (3)**, and the single **Sphenoid Bone (4)** are present. (The Occipital and Ethmoid bones are not evident in this anterior view). The **Coronal Suture (5)** is situated between the Frontal bone and the two Parietal bones posteriorly, while the **Sagittal Suture (6)** unites the two Parietal Bones in the midline. A small portion of the **Squamosal** (or Parieto-Temporal) **Suture (7)** is also seen. The bony **Orbital Cavity** surrounds the soft tissues of the Eye, with boundaries formed by the Frontal bone above, the Sphenoid and **Zygomatic (8)** bones laterally, the **Maxilla (9)** below, along with the **Nasal (10)** and **Lacrimal (11) Bones** on the medial side. Not clearly evident in this view are the Ethmoid Bones located posterior to the Lacrimal Bones, which also contribute to the formation of the Orbital Cavity. The opening of each Nasal Cavity is bounded by the Maxillary and Nasal Bones. Projecting from the lateral wall of the cavity toward the midline are three delicate paired folds of bone known as the **Nasal Conchae**, which have a scrolled or sea shell-like appearance. The Superior and Middle Conchae are projections of the Ethmoid bone, while the two large **Inferior Nasal Conchae (12)** are independent Facial bones. In the midline, is the **Bony Nasal Septum** dividing the two Nasal Cavities. The Septum is formed from the **Perpendicular Plate of the Ethmoid (13)** bone above and the **Vomer (14)** below. The **Body of the Mandible (15)** is also clearly evident in this view, along with the **Mastoid Process (16)** of the Temporal bones extending down on either side.

(1) Frontal Bone
(2) Parietal Bones
(3) Temporal Bones
(4) Sphenoid Bone
(5) Coronal Suture
(6) Sagittal Suture
(7) Squamosal Suture
(8) Zygomatic Bones
(9) Maxilla
(10) Nasal Bone
(11) Lacrimal Bone
(12) Inferior Nasal Concha
(13) Perpendicular Plate of the Ethmoid
(14) Vomer
(15) Body of the Mandible
(16) Mastoid Process of the Temporal

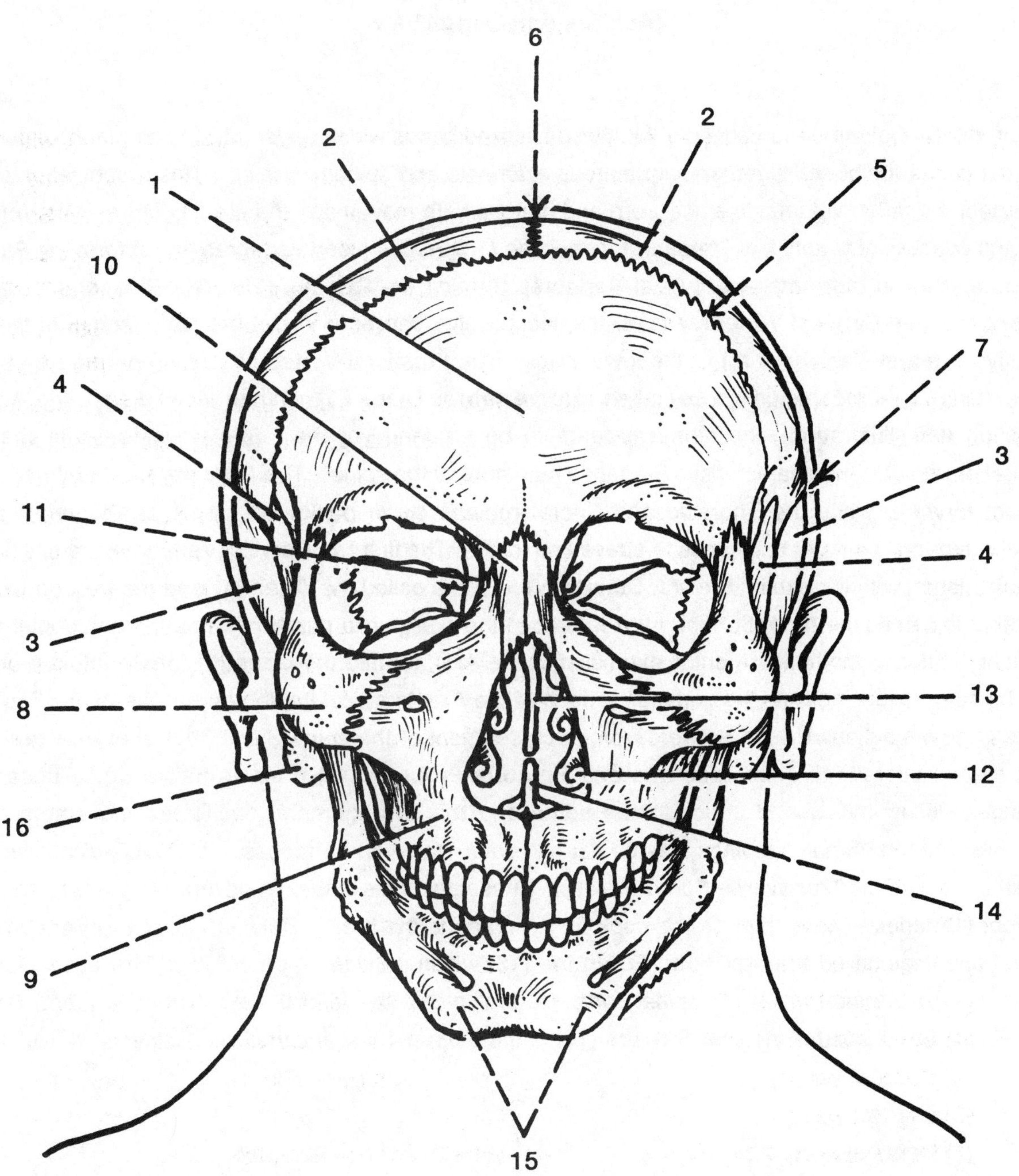
6
2
2
1
5
10
7
4
3
11
4
3
8
13
12
16
14
9
15

Vertebral Column

(Anterior and Lateral Views)

The Vertebral Column is composed of highly specialized bones which are separated by plates of cartilage and the bones are bound together by numerous ligaments and spinal muscles. This organization serves to protect the rather vulnerable spinal cord and permits both movement and erect posture. The Vertebral Column consists of twenty four **Presacral Vertebrae (1,2,3)**, five fused vertebrae which form the **Sacrum (4)**, and three to four small Coccygeal Vertebrae forming the **Coccyx (5)**. The Presacral Vertebrae consist of seven **Cervical Vertebrae (1)** of the neck, twelve **Thoracic Vertebrae (2)** attached to the ribs, and five **Lumbar Vertebrae (3)** of the lower back. The Presacral Vertebrae are separated by twenty three pads of Fibrocartilage tissue called **Intervertebral Discs (6)**. When viewed from the Anterior direction (left illustration), the spine appears to be straight, yet the lateral view (seen in the right illustration) reveals the characteristic S-shaped curvature of the spine. The Cervical and Lumbar regions project forward, while the Thoracic and Sacral regions curve backward. This arrangement of the vertebrae permits greater resistance to stress and force. The first two Cervical Vertebrae are specialized for articulation with the skull. The first Cervical Vertebra is called the **Atlas (7)** and the second Cervical Vertebra is named the **Axis (8)**. The Atlas is fixed to the skull, while the Axis forms the pivot point which permits rotation of the head. A small extension of the Axis, termed the **Dens (9)** (or Odontoid process), can be seen extending from the superior surface of the Axis to articulate above with the Atlas. Typically, three bony projections extend outward from each segment of the spine. Two **Transverse Processes (10)** extend laterally to either side and one **Spinous Process (11)** projects backward. These bony processes allow muscles to attach to the spine and this arrangement facilitates movement of the vertebral column. Small openings may be seen in each Transverse Processes of the Cervical Vertebrae which are called the **Transverse Foramina (12)**. These openings in the Transverse Processes allow the Vertebral Arteries to pass through the neck on their way to the brain. Between all of the Vertebrae are lateral openings called **Intervertebral Foramina (13)** which provide an unrestricted passageway for the spinal nerves to enter or exit the spinal cord. Note also that the lateral view of the Vertebral Column reveals the ear-shaped **Auricular Surface (14)** of the Sacrum that joins the Pelvis on each side.

(1) Cervical Vertebrae
(2) Thoracic Vertebrae
(3) Lumbar Vertebrae
(4) Sacrum
(5) Coccyx
(6) Intervertebral Discs
(7) Atlas
(8) Axis
(9) Dens of the Axis
(10) Transverse Processes
(11) Spinous Processes
(12) Transverse Foramina Foramen
(13) Intervertebral Foramina Foramen
(14) Auricular Surface of Sacrum

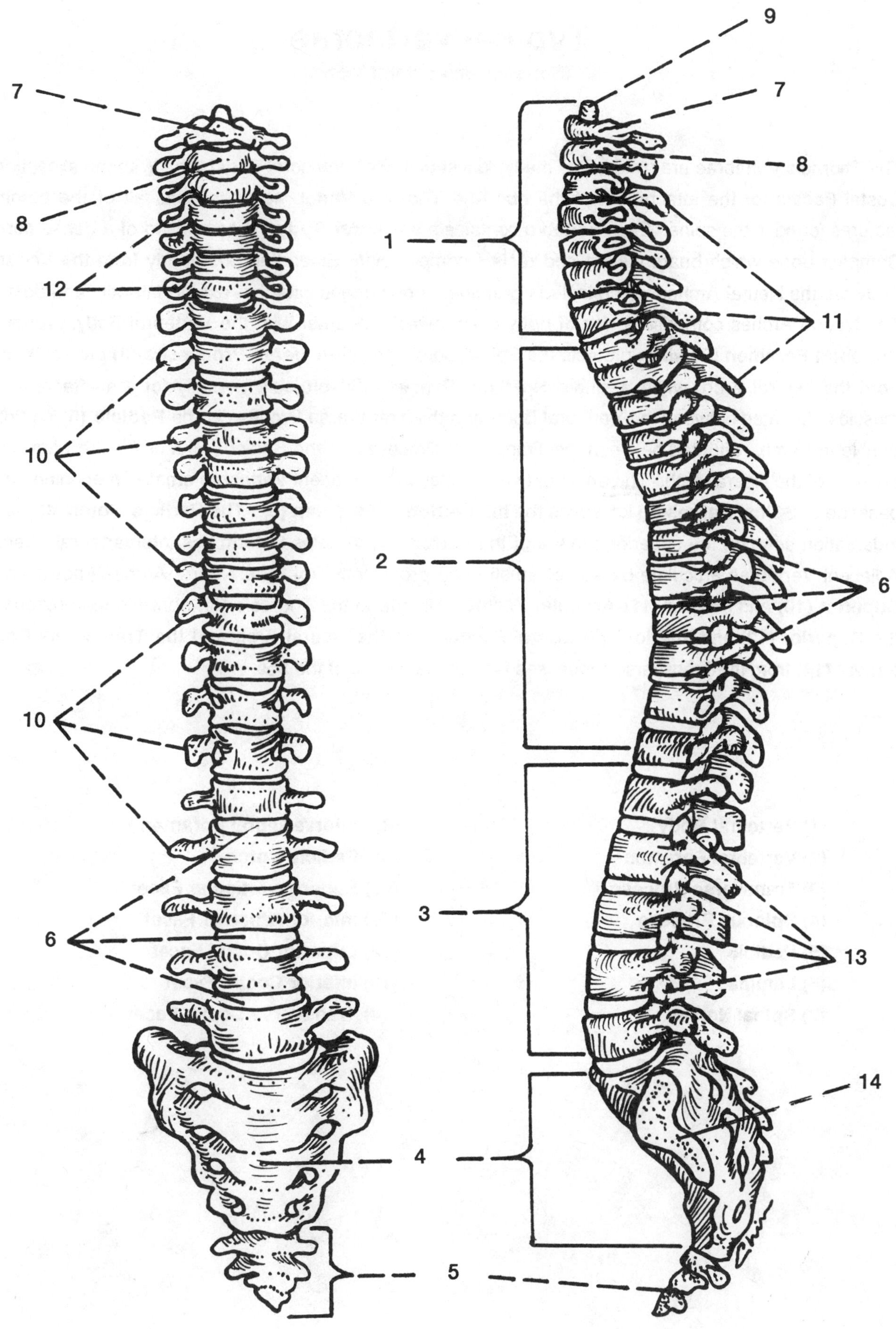

SS-5

Typical Vertebrae

(Superior and Lateral Views)

The Thoracic Vertebrae are pictured as the typical structure of Presacral Vertebrae. With the exception of Costal Facets for the attachment of the ribs, the Thoracic Vertebrae possesses all of the common features found in the spine. Each Vertebra contains a **Vertebral Body (1)** comprised of a dense core of Compact bone which bears weight and resists compression. Extending posteriorly from the Vertebral Body are the Neural Arches which include prominent bony projections that extend laterally and dorsally. The Neural Arches comprise a ring of bony tissue which, together with the Vertebral Body, create the **Vertebral Foramen (2)** which protects the Spinal Cord. Two **Transverse Processes (3)** project laterally from the Neural Arch, while a single **Spinous Process (4)** projects dorsally for the attachment of muscles. Located between the Vertebral Body and the Transverse Process is the **Pedicle (5)** portion of the Neural Arch. Situated between the Transverse Process and the Spinous Process is the **Lamina (6)** element of the Neural Arch. The union of two Pedicles from adjacent Vertebrae create an opening for the passage of **Spinal Nerves (7)** known as the **Intervertebral Foramen (8)**. The **Pedicle Notch (9)** is the indentation seen on the superior surface of the Pedicle which helps to form this Intervertebral opening. Adjacent Vertebrae articulate by way of small bony projections from the Neural Arches known as the **Superior (10)** and **Inferior (11) Articular Facets**. Unique to the Thoracic Vertebrae is the presence of the **Superior (12)** and **Inferior (13) Costal Facets** from the Neural Arch and the **Transverse Costal Facets (14)** from the Transverse Processes for the attachment of the ribs.

(1) Vertebral Body
(2) Vertebral Foramen
(3) Transverse Process
(4) Spinous Process
(5) Pedicle
(6) Lamina
(7) Spinal Nerves
(8) Intervertebral Foramen
(9) Pedicle Notch
(10) Superior Articular Facet
(11) Inferior Articular Facet
(12) Superior Costal Facet
(13) Inferior Costal Facet
(14) Transverse Costal Facet

Ribs

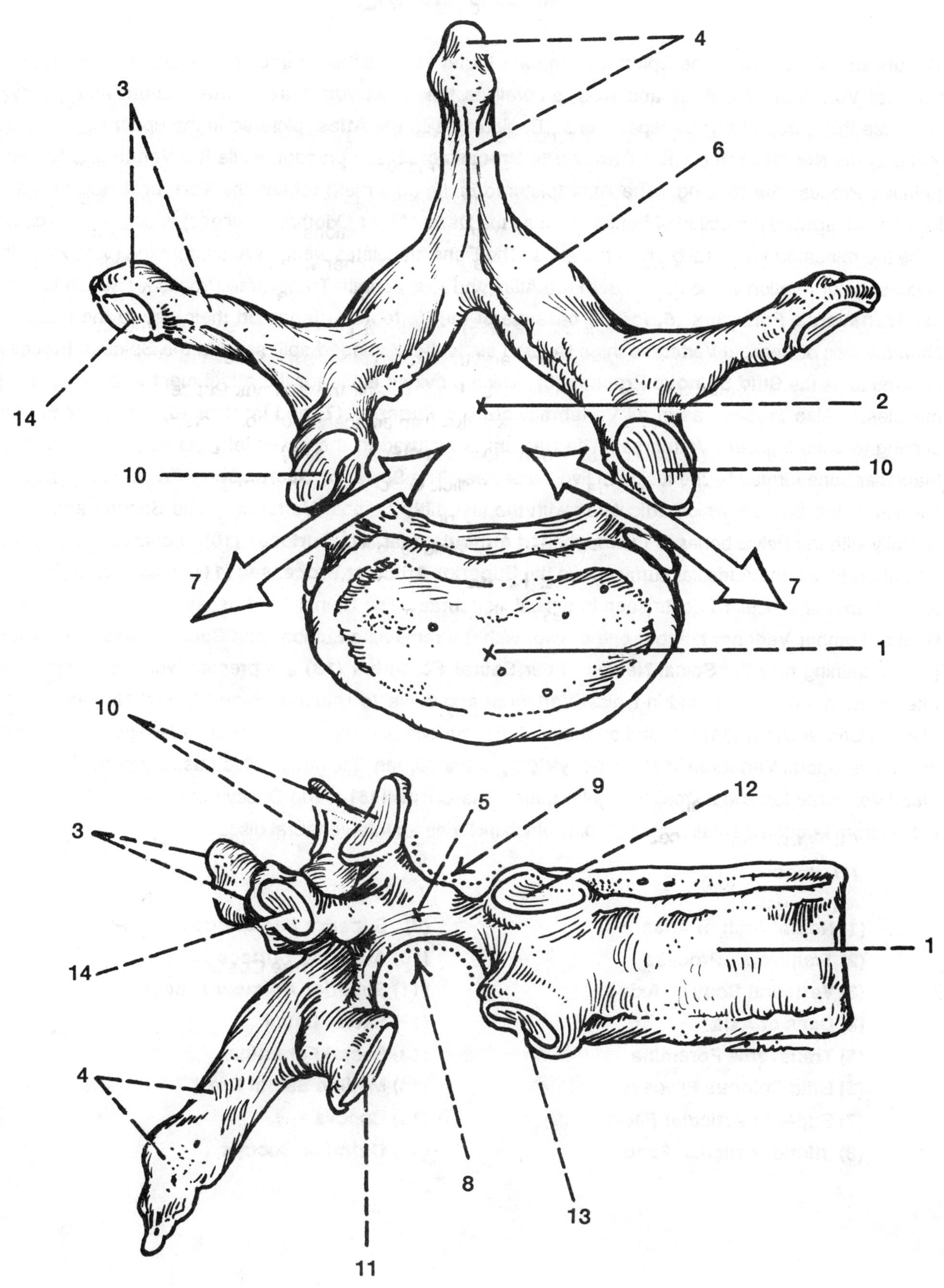
4
3
6
14
2
10
10
7
7
1
10
9
12
5
3
14
1
4
8
13
11

Atypical Vertebrae

(Posterior Views)

The unusual Vertebrae of the Spinal Column are found at the extreme ends of the chain. The first two Cervical Vertebrae, the Atlas and Axis, as well as the fused vertebrae of the Sacrum and Coccyx, comprise this group of atypical specimens. Beginning with the **Atlas** (pictured in the upper right), notice that only the **Neural Arch (1)** and **Transverse Processes (2)** are present, while the Vertebral Body and Spinous Process are missing. The **Axis** (pictured in the upper left) retains the **Vertebral Body (3)** and includes an upward projection of bone known as the **Dens (4)** (or Odontoid process) which is considered to be the relocated Vertebral body of the Atlas. The Dens articulates with the Atlas, forming the pivot point necessary for rotation of the head. Both the Atlas and Axis contain Transverse Processes which include the **Transverse Foramina (5)** for the passage of the Vertebral Arteries on their way to the brain. A characteristic of Cervical Vertebrae three through six is the V-shaped split seen in the Spinous Process, referred to as the **Bifid Spinous Process (6)**, which provides a site for the attachment of powerful neck muscles. Also present, as in all Vertebrae, are the **Superior (7)** and **Inferior (8) Articular Facets** serving to unite adjacent Vertebrae. The Sacrum is pictured in the lower left and consists of a large triangular bone formed by the fusion of five Vertebrae. The **Superior Sacral Body (9)** marks the upper surface of the Sacrum which articulates with the last (fifth) Lumbar Vertebra. The Sacrum articulates laterally with the Pelvic bones at the roughened **Auricular** (ear-like) **Surfaces (10)**. Located between the Sacral Body and the Auricular surfaces are the **Superior Articular Processes (11)**, which are equivalent to the Superior Articular Facets seen in typical Vertebrae. The Spinal Cord usually extends only to the level of Lumbar Vertebra number one or two, with the remaining Lumbar and **Sacral Vertebral Canals (12)** containing only the Spinal Nerves. Four **Sacral Foramina (13)** are present which correspond to Intervertebral Foramina found in typical Vertebrae and serve to transmit Sacral Spinal Nerves. The **Median Sacral Crest (14)** is found on the posterior surface and represents the fused Spinous Processes of the five Sacral Vertebrae. The **Coccyx (15)** is the human "tail bone" which usually consists of four fused Vertebrae forming a small triangular bone. The **Cornu (16)** of the Coccyx is united to the Apex of the Sacrum through a small Synovial Joint which includes an Intervertebral disc.

(Vertebral)

(1) Neural Arch of Atlas – Omit
(2) Transverse Process
(3) Vertebral Body of Axis
(4) Dens of Axis
(5) Transverse Foramina Foramen
(6) Bifid Spinous Process
(7) Superior Articular Facet
(8) Inferior Articular Facet

(9) Superior Sacral Body
(10) Auricular Surface
(11) Superior Articular Process
(12) Sacral Canal
(13) Sacral Foramina
(14) Median Sacral Crest
(15) Coccyx
(16) Cornu of Coccyx

Omit

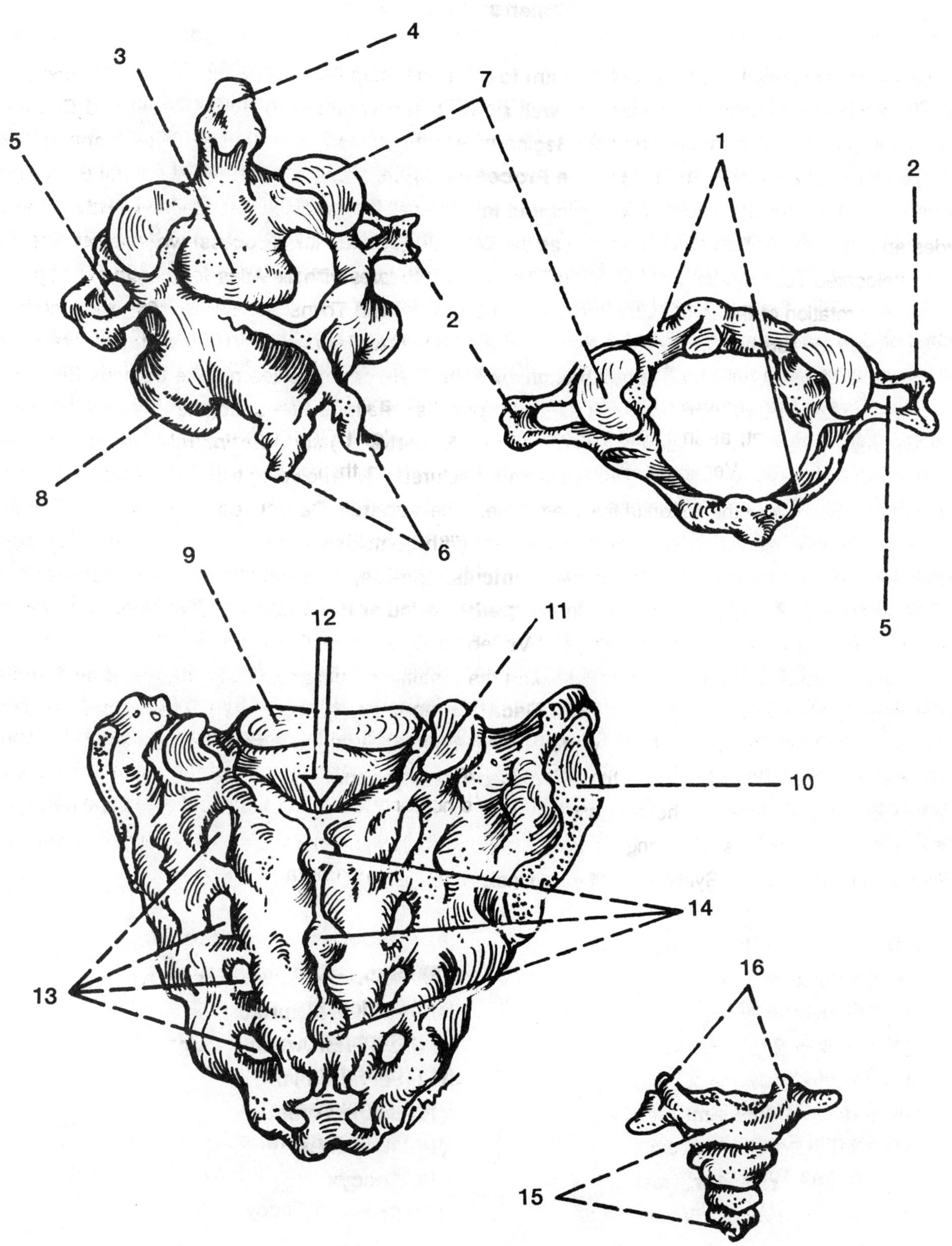
4
3
7
5
1
2
2
8
6
5
9
12
11
10
14
13
16
15

Ribs and Sternum

The **Thoracic Skeleton** includes the twelve pairs of Ribs, twelve Thoracic Vertebrae, the Sternum, two Clavicles, and two Scapulae. Pictured here is a typical rib represented by the Sixth rib seen on the lower right, and one example of a rib with some unique properties represented by the First rib seen on the upper left. Beginning with the Sixth rib, notice that each end of the rib is specialized to permit articulation with either the Vertebral Column posteriorly or with the Sternum anteriorly. The Vertebral end displays two **Articular Facets (1)** which form joints with the Bodies of two adjacent Thoracic Vertebrae. The Vertebral portion of the rib also includes the expanded **Head (2)**, the more narrow **Neck (3)**, and the **Tubercle (4)** which forms a joint with the Transverse Process of the nearby Thoracic Vertebra. As the rib sweeps forward and laterally, there is a sudden bend toward the midline which is referred to as the **Rib Angle (5)**. The Inferior Surface of the rib includes a narrow **Costal Groove (6)** for the **Costal Arteries and Veins** which travel under the protection of the rib cage. The **Sternal Extremity (7)** of the rib extends anteriorly to join with a **Costal Cartilage** which forms a movable joint with the Sternum. The First Rib, like the Sixth rib, also contains a **Head, Neck (3)**, and **Tubercle** for articulation with the Thoracic vertebrae. Note that the Head of this First Rib contains a single **Articular Facet (1)** for articulation with the Body of the First Thoracic Vertebra. The First Rib displays the greatest curvature of all the ribs and includes the **Scalene Tubercle (8)** for attached of the **Anterior Scalene** neck muscle. Turning to the Sternum, notice that it is composed of three bones: the uppermost expanded **Manubrium (9)**, the middle elongated **Sternal Body (10)**, and the delicate **Xiphoid Process (11)** below. Located between the Manubrium and Sternal Body is the **Sternal Angle (12)** where the Second Ribs articulate on each side. The **Xiphisternal Junction (13)** marks the union of the Xiphoid Process with the Sternal Body, and the location where the Seventh Ribs join the Sternum. On each side of the Manubrium is a **Clavicular Notch** (or Facet) **(14)** which forms a Synovial Joint with the medial end of the Clavicle. The Sternal Extremity of the First Rib also articulates with the Manubrium at the **First Costal Notch (15)**. Along the Superior margin of the Manubrium is the **Suprasternal** (or Jugular) **Notch (16)** which is bridged by the Interclavicular Ligament extending between the two Clavicles.

(1) Articular Facet
(2) Head of the Rib
(3) Neck of the Rib
(4) Tubercle of the Rib
(5) Rib Angle
(6) Costal Groove
(7) Sternal Extremity
(8) Scalene Tubercle

(9) Manubrium of the Sternum
(10) Body of the Sternum
(11) Xiphoid Process
(12) Sternal Angle
(13) Xiphisternal Junction
(14) Clavicular Notch
(15) First Costal Notch
(16) Suprasternal Notch

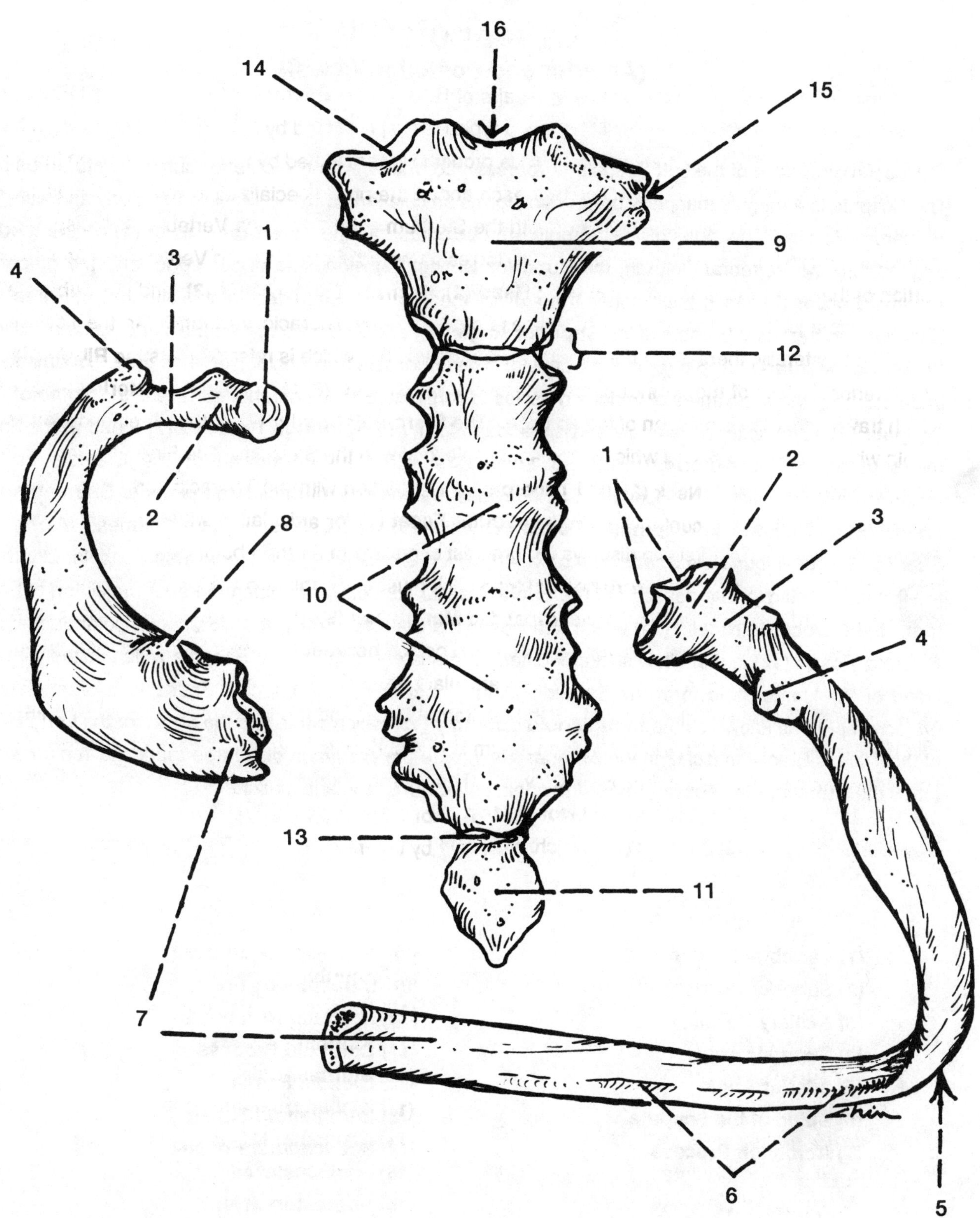

SS-8

Scapula
(Anterior and Posterior Views)

The posterior surface of the Left Scapula is represented above, with the anterior surface pictured below. The Scapula is a roughly triangular bone which forms articulations with the Clavicle anteriorly and with the Humerus laterally. The sides of this triangle are designated as the **Vertebral Border (1)** on the medial edge facing the Vertebral Column, the **Superior Border (2)** along the upper edge, and the diagonal **Axillary Border (3)** which approaches the Axillary fossa of the Shoulder Joint. Situated between the Superior and Vertebral Borders is the **Superior Angle (4)**, and located between the Vertebral and Axillary Borders is the **Inferior Angle (5)** of the Scapula. These Borders and Angles serve as useful landmarks in understanding the attachment of various muscles. Prominent on the posterior surface is the **Spine of the Scapula (6)** which extends laterally and expands to form the **Acromion Process (7)** found at the tip of the shoulder. The Acromion Process articulates with the lateral end of the Clavicle at the Acromioclavicular Joint. The Scapular Spine divides the posterior surface into the upper **Supraspinous Fossa (8)** and a lower **Infraspinous Fossa (9)** which house muscles of similar names. Along the Superior Border is found the **Scapular Notch (10)** which is bridged by the Transverse Scapular Ligament to form an opening for the passage of the Suprascapular Nerve. A prominent anterior projection of bone forms the **Corocoid Process (11)** which attaches to the Clavicle forming the Coracoclavicular Joint. The **Glenoid Fossa (12)** is a shallow socket located on the lateral edge of the Scapula which unites with the Head of the Humerus to form the Shoulder Joint. Just below the Glenoid Fossa is a sharp lateral projection of bone known as the **Infraglenoid Tubercle (13)** which serves as the origin for the Long Head of the Triceps Brachii muscle. In the Anterior View, note the presence of a large triangular region which forms the **Subscapular Fossa (14)** for the origin of the Subscapularis muscle.

(1) Vertebral Border
(2) Superior Border
(3) Axillary Border
(4) Superior Angle
(5) Inferior Angle
(6) Spine of the Scapula
(7) Acromion Process
(8) Supraspinous Fossa
(9) Infraspinous Fossa
(10) Scapular Notch
(11) Corocoid Process
(12) Glenoid Fossa
(13) Infraglenoid Tubercle
(14) Subscapular Fossa

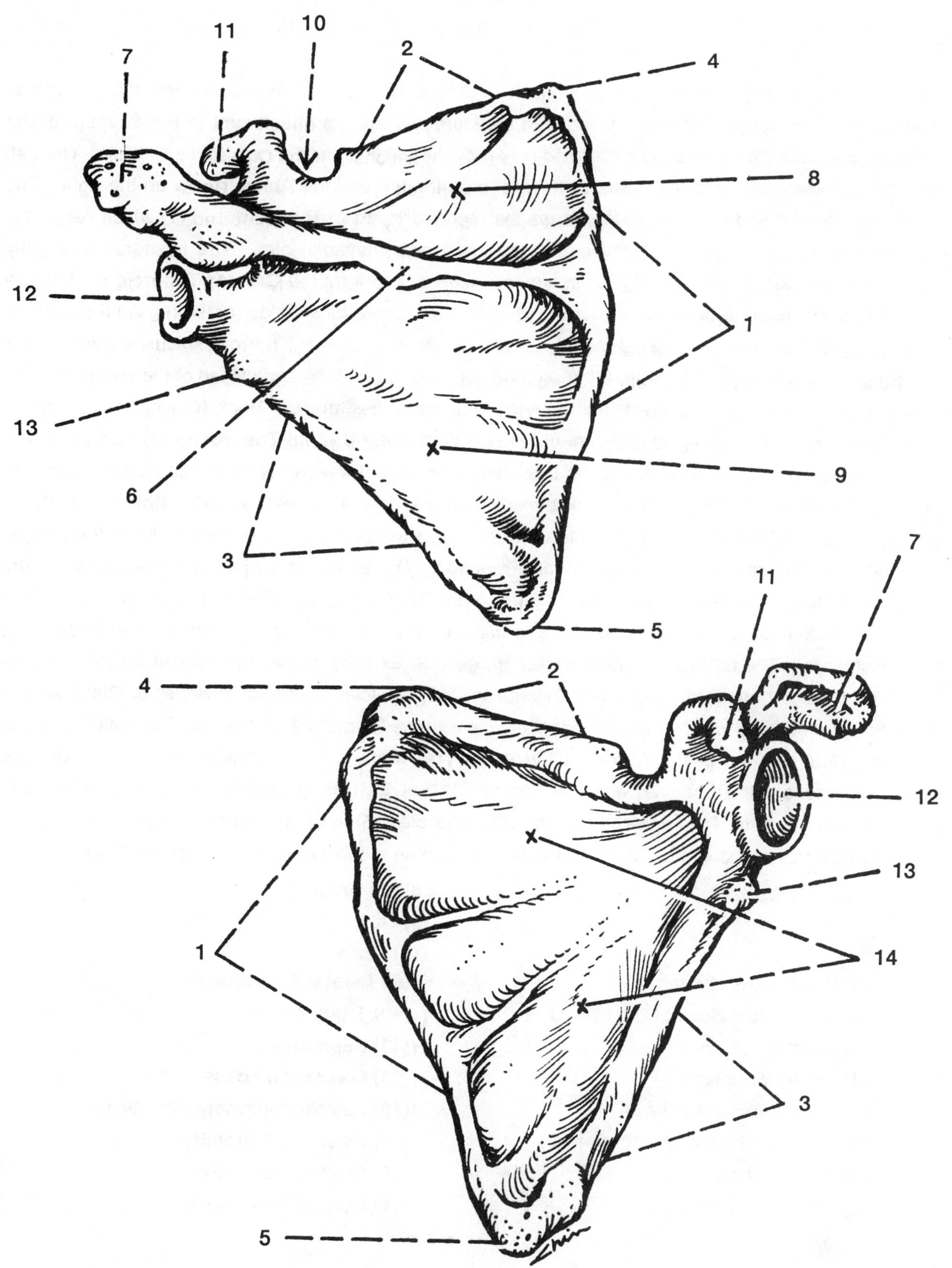

SS-9

Humerus and Clavicle

The Upper and Lower Extremities each consist of sixty bones, with some similarities and yet unique differences. The Upper Extremity begins with the Humerus and its attachment to the Scapula at the Shoulder joint, with the top of this joint formed in part by the union of the Scapula to the Clavicle. The Left Humerus is shown here, with the Posterior view seen on the left and the Anterior view on the right. The expanded proximal end of the Humerus forms the rounded **Head of the Humerus (1)** which lies in the Glenoid Fossa of the Scapula, forming a loose and freely movable joint. The Humerus is slightly narrowed at the **Anatomical Neck (2)**, yet widens below this to form the **Greater Tubercle (3)** laterally and the **Lesser Tubercle (4)** anteriorly for the attachment of powerful shoulder muscles. On the Anterior surface situated between the Greater and Lesser Tubercles is a long, narrow sulcus known as the **Intertubercular** (or Bicipital) **Groove (5)** which houses the tendon of the Long Head of the Biceps Brachii muscle. Below the Tubercles, the Humerus narrows to form the **Surgical Neck (6)** which is a frequent site of fractures. Just above the midpoint of the Humerus is the **Deltoid Tuberosity (7)** located on the ventrolateral surface for the insertion of the Deltoid muscle. The distal end of the Humerus is expanded to form the articulations for the elbow joint. The medial distal surface includes the **Medial Epicondyle (8)** for the origin of several flexor muscles of the forearm and hand, while the lateral surface includes the **Lateral Epicondyle (9)** for the origin of several extensor muscles. The actual elbow joint is formed by the more medial **Trochlea** (a "pulley") **(10)** joining the Ulna, while the more lateral **Capitulum** ("little head") **(11)** joining the Radius and the expanded upper end of the Ulna articulates within the **Olecranon Fossa (12)** of the Humerus. The **Lateral Supracondylar Ridge (13)** extends above the Lateral Epicondyle and serves as the bony origin for the powerful Brachioradialis muscle. The "Collar Bone" or **Clavicle** is a rather flat, curved bone which forms important unions with the Sternum and Scapula. The Left Clavicle is pictured. Note that the lateral **Acromial Extremity (14)** forms the Acromioclavicular Joint with the Acromial Process of the Scapula, while the medial **Sternal Extremity (15)** forms the Sternoclavicular Joint at the Manubrium. The cone-shaped **Conoid Tubercle (16)** is located on the inferior surface of the Clavicle and attaches the strong Coracoclavicular Ligament to the Coracoid Process of the Scapula.

(1) Head of the Humerus
(2) Anatomical Neck of the Humerus
(3) Greater Tubercle
(4) Lesser Tubercle
(5) Intertubercular Groove
(6) Surgical Neck of the Humerus
(7) Deltoid Tuberosity
(8) Medial Epicondyle
(9) Lateral Epicondyle
(10) Trochlea
(11) Capitulum
(12) Olecranon Fossa
(13) Lateral Supracondylar Ridge
(14) Acromial Extremity
(15) Sternal Extremity
(16) Conoid Tubercle

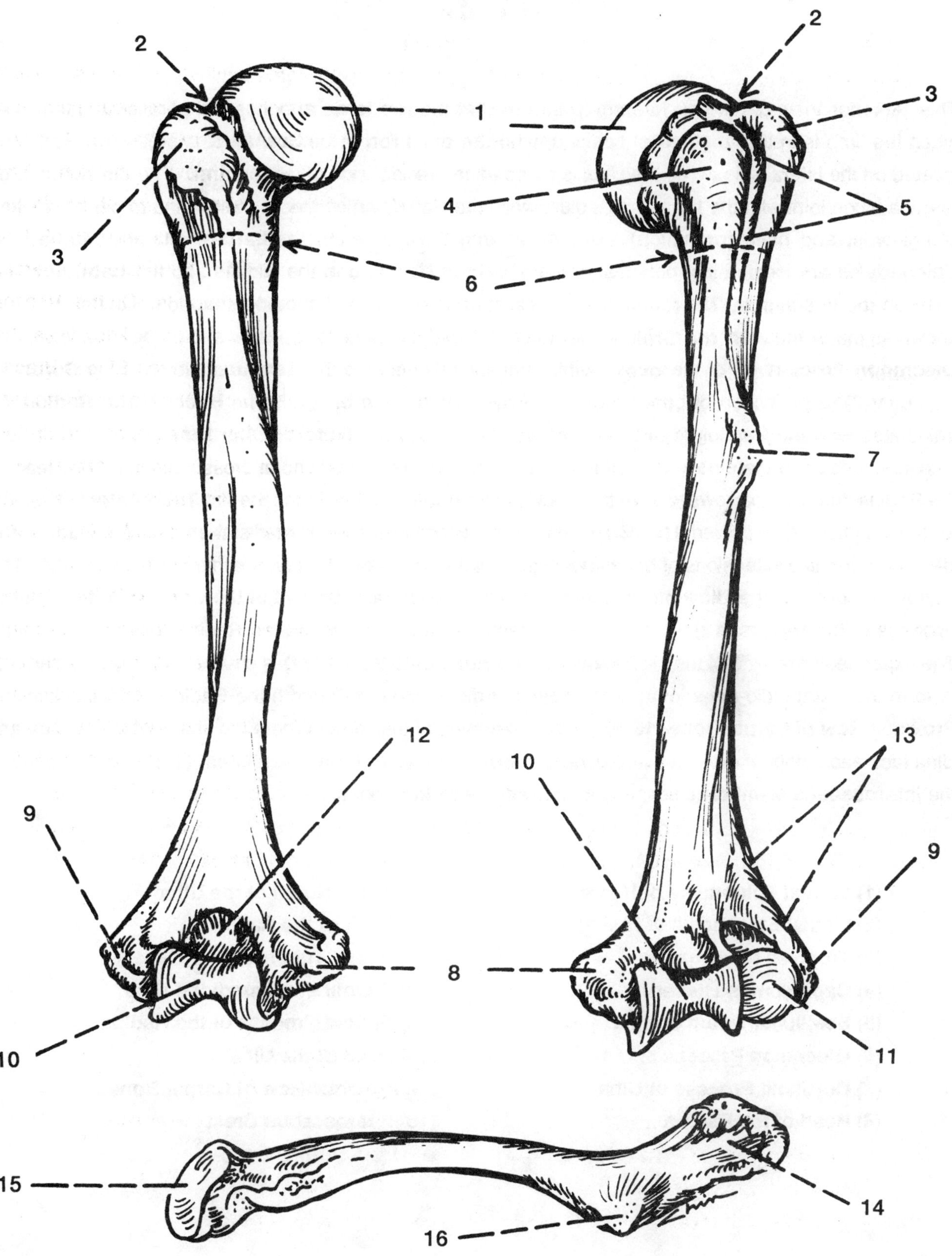
2
1
3
4
5
6
7
8
9
10
11
12
13
14
15
16

Forearm Bones: Radius and Ulna

(Anterior View)

This Anterior View of the Left forearm illustrates that the two forearm bones are parallel to each other when the limb is in the anatomical position, with the palm facing upward (supinated). The Radius is located on the lateral side, while the Ulna is found on the medial side of the forearm. The distal Humerus forms a hinge joint with the Ulna and Radius, while the Radius articulates with the Ulna to permit rotation of the wrist and hand (pronation). Beginning with the Humerus, the **Lateral (1)** and **Medial (2) Epicondyles** are located just proximal to the **Trochlea (3)** found in the midline and the **Capitulum (4)** seen on the lateral side. The rounded Trochlea articulates with a half moon-shaped hollow of the Ulna known as the **Semilunar** (or Trochlear) **Notch (5)**, while the expanded proximal end of the Ulna forms the **Olecranon Process** (6) of the Elbow Joint. Immediately distal to the Semilunar Notch is the **Coronoid Process (7)**, a projection of bone where the Brachialis muscle inserts. The **Head of the Radius (8)** articulates with the Capitulum of the Humerus and the **Radial Notch of the Ulna (9)** to permit both flexion/extension of the elbow and rotation of the wrist as the forearm bone cross. Distal to the Head of the Radius, the shaft narrows to form the **Neck of the Radius (10)** with the **Radial Tuberosity (11)** below for the insertion of the powerful tendon of the Biceps Brachii muscle. The distal ends of the Radius and Ulna form pen-like extensions of bone known as Styloid Processes, found on either side of the wrist. The **Styloid Process of the Ulna (12)** attaches the Ulnar Collateral Ligament of the wrist, while the **Styloid Process of the Radius (13)** attaches to the Brachioradialis muscle and the Radial Collateral Ligament. The expanded lower end of the Ulna known as the **Head of the Ulna (14)** is covered with hyaline cartlage to form the distal radio-ulnar joint. The expanded distal ends of the both the Radius and Ulna join the **Proximal Row of Carpal Bone (15)** to permit movement of the wrist. Where the shafts of the Radius and Ulna face each other, there is a sharp edge of bone called the **Interosseous Crest (16)** which attaches to the **Interosseous Membrane** suspended between these two bones.

(1) Lateral Epicondyle of Humerus
(2) Medial Epicondyle of Humerus
(3) Trochlea of Humerus
(4) Capitulum of Humerus
(5) Semilunar Notch of Ulna
(6) Olecranon Process of Ulna
(7) Coronoid Process of Ulna
(8) Head of the Radius
(9) Radial Notch of the Ulna
(10) Neck of the Radius
(11) Radial Tuberosity
(12) Styloid Process of the Ulna
(13) Styloid Process of the Radius
(14) Head of the Ulna
(I5) Proximal Row of Carpal Bones
(16) Interosseous Crest

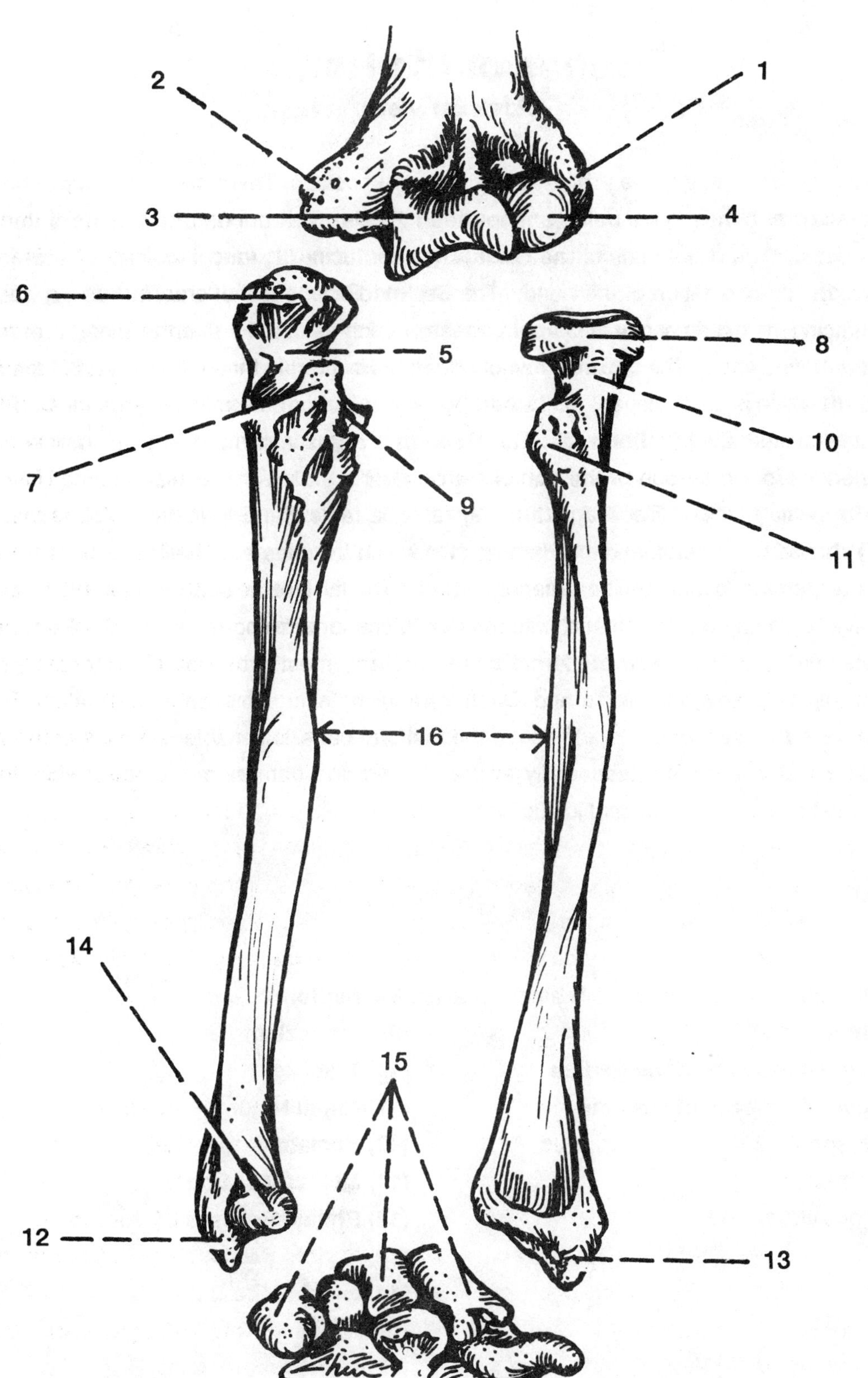
1
2
3
4
6
5
8
10
7
9
11
16
14
15
12
13

Bones of the Hand

(Posterior View)

The Left Hand is pictured here in the Pronated (Palm down) position. There are eight **Carpal** bones of the wrist, five **Metacarpal** bones in the palm, and fourteen **Phalangial** bones of the fingers, for a total of twenty-seven bones. The distal ends of the **Radius (1)** and **Ulna (2)** form the Radio-Ulnar joint which permits pronation and supination of the hand. The **Styloid Processes of the Radius (3)** and **Ulna (4)** are seen embracing the proximal row of Carpal bones on each side. The Carpal Bones are arranged in two rows of four bones each. The proximal row of bones (from lateral to medial) consists of the **Scaphoid** (or Navicular) **(5)** which is boat-shaped, the **Lunate (6)** which is moon-shaped, the triangular **Triquetrum (7)**, and the small pea-shaped **Pisiform (8)**. The Pisiform is an example of a Sesamoid bone because it is found suspended in the tendon of the Flexor Carpi Ulnaris muscle. The distal row of Carpal bones (from lateral to medial) include the **Trapezium (9)** (a "little table"), the four-sided **Trapezoid (10)**, the **Capitate (11)** shaped like a head, and the **Hamate (12)** which includes a hook-like process known as the **Hamulus of the Hamate** located on the Anterior surface. The five **Metacarpal Bones (13)** are numbered one through five beginning with the thumb, with the First Metacarpal being the shortest. Each Metacarpal bone articulates with the Distal row of Carpal bones and the most proximal of the Phalangial bones. Three **Phalanges (14)** (proximal, middle and distal) form each finger, while only two Phalanges (proximal and distal) are found in the thumb. In addition to the Pisiform bone, a variable number of small sesamoid bones may be found in the hand, especially at the second and fifth metacarpo-phalangeal joints and occasionally among the interphalangeal joints.

(1) Radius
(2) Ulna
(3) Styloid Process of the Radius
(4) Styloid Process of the Ulna
(5) Scaphoid
(6) Lunate
(7) Triquetrum
(8) Pisiform
(9) Trapezium
(10) Trapezoid
(11) Capitate
(12) Hamate
(13) Metacarpal Bones
(14) Phalanges

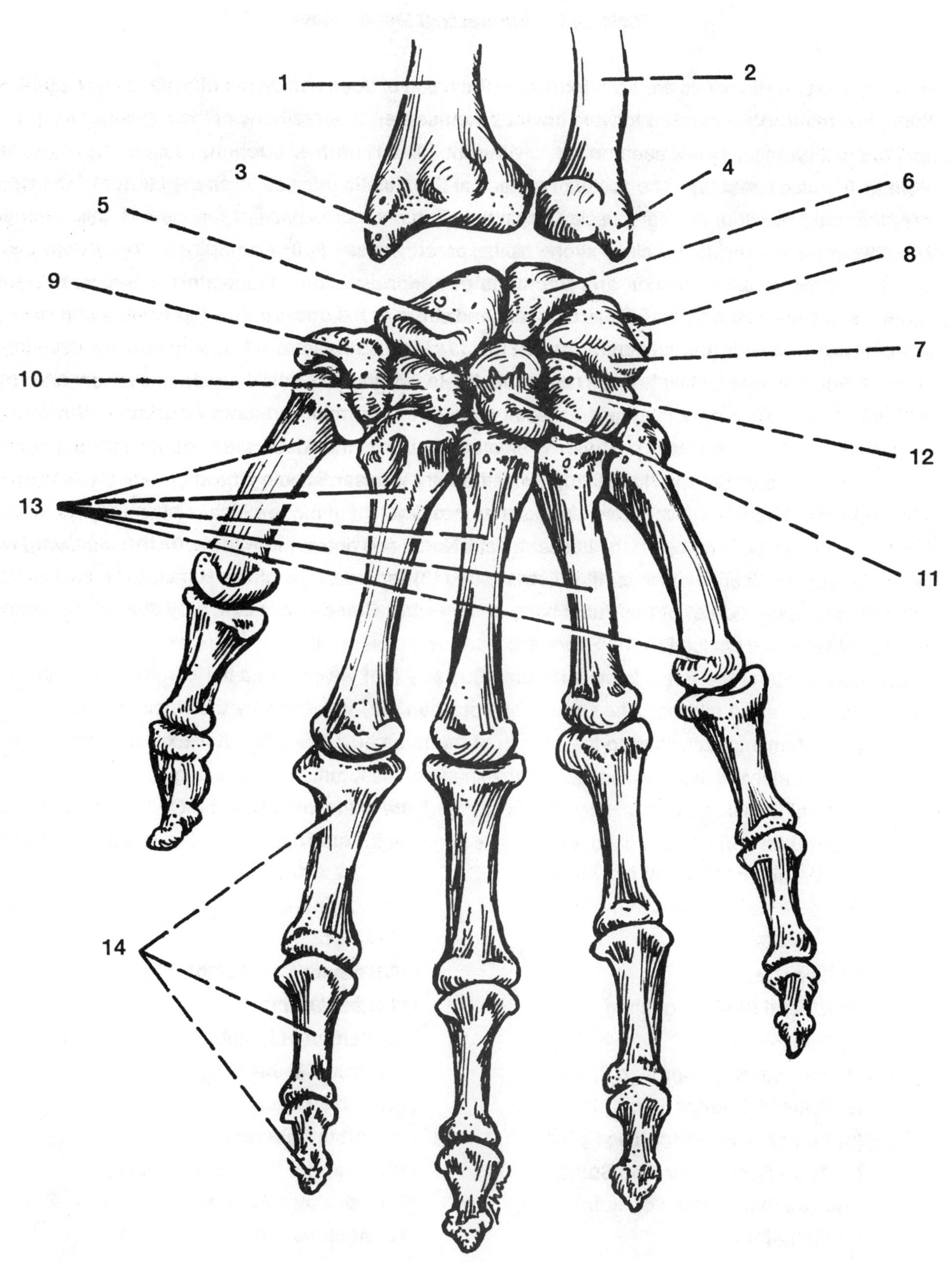

SS-12

Hip Bone

(Right Side, Lateral and Medial Views)

The Right and Left Hip bones are each formed by the union of three bones: the **Ilium** above, the **Ischium** behind, and the **Pubis** extending forward towards the midline. A lateral view of the Hip is pictured on the upper left, with a medial view seen on the lower right. The Ilium has a rough, rounded upper surface known as the **Iliac Crest (1)**. The posterior surface of the Ilium is referred to as the **Gluteal Surface (2)** where the three powerful Gluteal muscles originate. The anterior surface of the Ilium is called the **Iliac Fossa (3)** where the Iliacus muscle, a strong hip flexor, originates. Four Iliac Spines project from the hip bone, two in the anterior direction and two in the posterior direction. The **Anterior Superior (4)** and **Anterior Inferior (5) Iliac Spines** project forward and serve as the origin of two hip flexor muscles, while the **Posterior Superior (6)** and **Posterior Inferior (7) Iliac Spines** project backward and attach to the Sacro-tuberous ligament. The **Sacrum** is situated between the two Pelvic bones posteriorly and forms the Sacro-iliac Joints on the roughened medial edge of the Ilium known as the **Auricular Surface (8)**. The Ischium includes a sharp posterior projection of bone known as the **Ischial Spine (9)** which separates the **Greater Sciatic Notch (10)** above from the **Lesser Sciatic Notch (11)** below. Ligaments attached to the Ischial Spine convert there notches into Sciatic Foramina which contain important nerves of the lower extremity. Just below the Lesser Sciatic Notch and above the **Ramus of the Ischium (12)** is the roughened prominence of the **Ischial Tuberosity (13)** for the origin of the Hamstring muscles of the thigh. On the upper border of the Pubic bone is the **Pubic Crest (14)**, marked by the **Pubic Tubercle (15)** laterally which is crossed by the Spermatic Cord as it passes through the Inguinal Canal of the male. On the medial side of the Pubis is the **Articular Surface (16)** which forms the cartilaginous joint of the Symphysis Pubis in the midline. The **Obturator Foramen (17)** is formed by the Pubis and Ischium with the Obturator Membrane attached to the margin of the foramen. The Obturator Externus muscle arises from the outer surface of this membrane, with the shape of the foramen being triangular in the female and oval in the male. The Ilium, Ischium, and Pubis all contribute to the formation of a C-shaped cavity known as the **Acetabulum (18)** which is lined with Hyaline caritlage for articulation with the **Head of the Femur** within the Hip Joint.

(1) Iliac Crest
(2) Gluteal Surface of Ilium
(3) Iliac Fossa
(4) Anterior Superior Iliac Spine
(5) Anterior Inferior Iliac Spine
(6) Posterior Superior Iliac Spine
(7) Posterior Inferior Iliac Spine
(8) Auricular Surface of Ilium
(9) Ischial Spine
(10) Greater Sciatic Notch
(11) Lesser Sciatic Notch
(12) Ramus of Ischium
(13) Ischial Tuberosity
(14) Pubic Crest
(15) Pubic Tubercle
(16) Articular Surface of Pubis
(17) Obturator Foramen
(18) Acetabulum

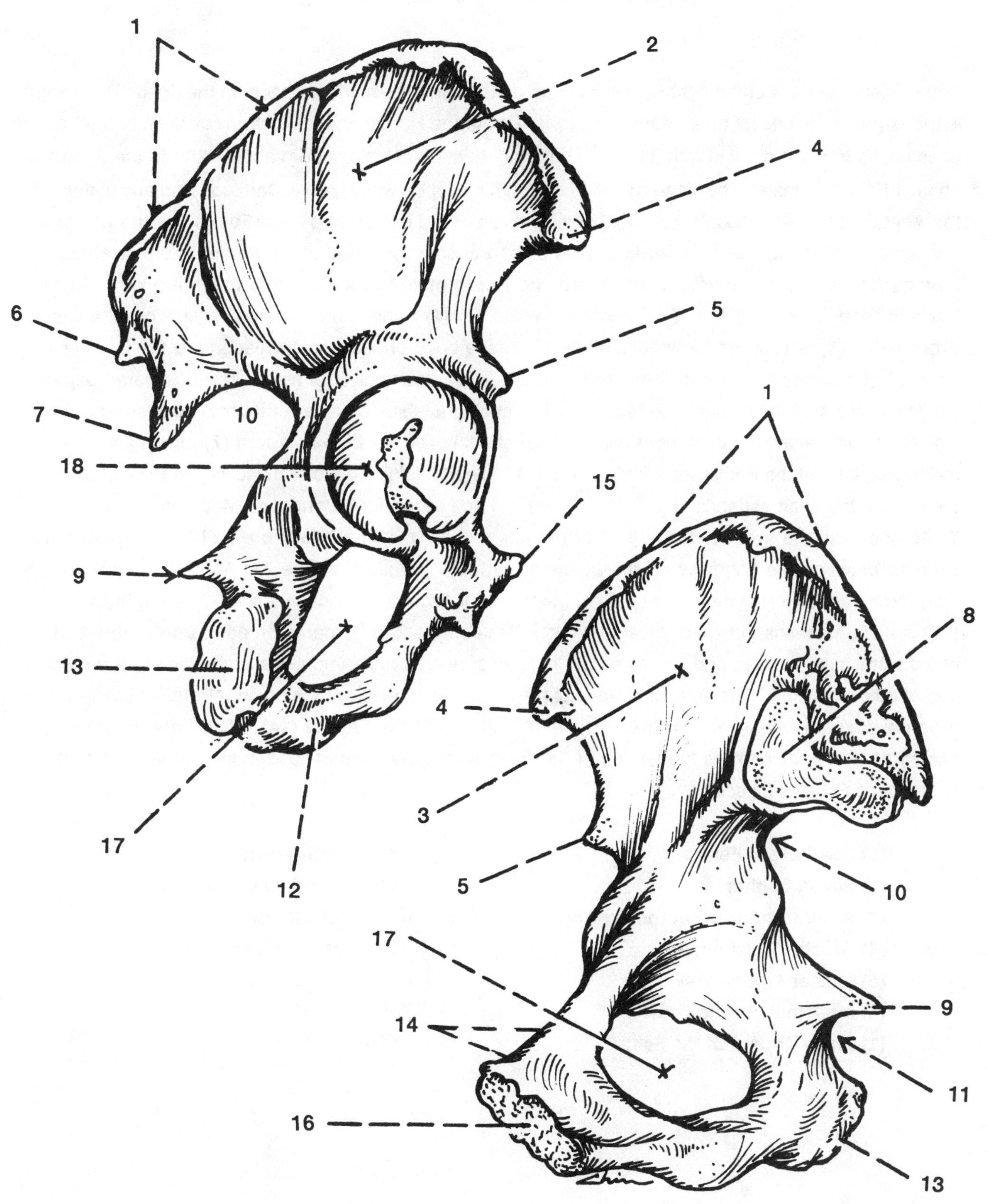

SS-13

Femur and Patella

(Right side, Anterior and Posterior Views)

The Posterior surface of the right Femur appears on the left, the Anterior surface on the right. The Femur is the largest and longest bone of the skeleton. Unlike the Humerus, the Anatomical Neck of the Femur does not lie in a straight line with the shaft, but at an angle that ranges from about 120° in the female to about 140° in the male. The **Head of the Femur (1)** is capped with Hyaline Cartilage and articulates with the **Acetabulum** of the Hip. On the medial side of the Head is a small pit, the **Fovea Capitis (2)** for the attachment of the Ligamentum Teres which helps to anchor the Head of the Femur in the Acetabulum. Beyond the Head, the Femur narrows to form the **Anatomical Neck (3)**. Note that there is no Surgical Neck as seen in the Humerus. The **Greater Trochanter (4)** is found on the lateral side, while the **Lesser Trochanter (5)** is seen on the medial surface. These bony projections correspond to the Greater and Lesser Tuberosities seen on the Humerus and serve to attach powerful hip muscles. Situated between the Trochanters on the posterior surface and descending at a steep angle is the **Intertrochanteric Crest (6)**. Continuing along the posterior surface of the shaft, notice the **Linea Aspera (7)** running vertically in the middle third of the Femur for the attachment of powerful thigh Adductor muscles. At the distal end of the Femur, the shaft expands to form the **Medial (8)** and **Lateral (9) Condyles** which articulate with the **Tibia** and form the synovial hinge joint of the knee. The **Intercondylar Fossa (10)** is a deep notch situated between the condyles on the posterior surface. Located above the Medial Condyle is the **Adductor Tubercle (11)** for the insertion of the powerful Adductor Magnus muscle. The **Popliteal Fossa (12)** is a diamond-shaped space located behind the knee which is bordered on each side by the tendons of the Hamstring muscles and contains arteries, veins, nerves and lymphatics embedded in a protective bed of adipose tissue. The anterior surface of the distal Femur displays the smooth **Patellar Surface (13)** which is covered with Hyaline Cartilage for articulation with the **Patella (14)**. The Patella (commonly known as the "knee cap") is the largest of the Sesamoid bones, found suspended in the tendon of the Quadriceps Femoris muscle.

(1) Head of the Femur
(2) Fovea Capitis
(3) Anatomical Neck of the Femur
(4) Greater Trochanter
(5) Lesser Trochanter
(6) Intertrochanteric Crest
(7) Linea Aspera of the Femur
(8) Medial Condyle of the Femur
(9) Lateral Condyle of the Femur
(10) Intercondylar Fossa
(11) Adductor Tubercle
(12) Popliteal Fossa
(13) Patellar Surface
(14) Patella

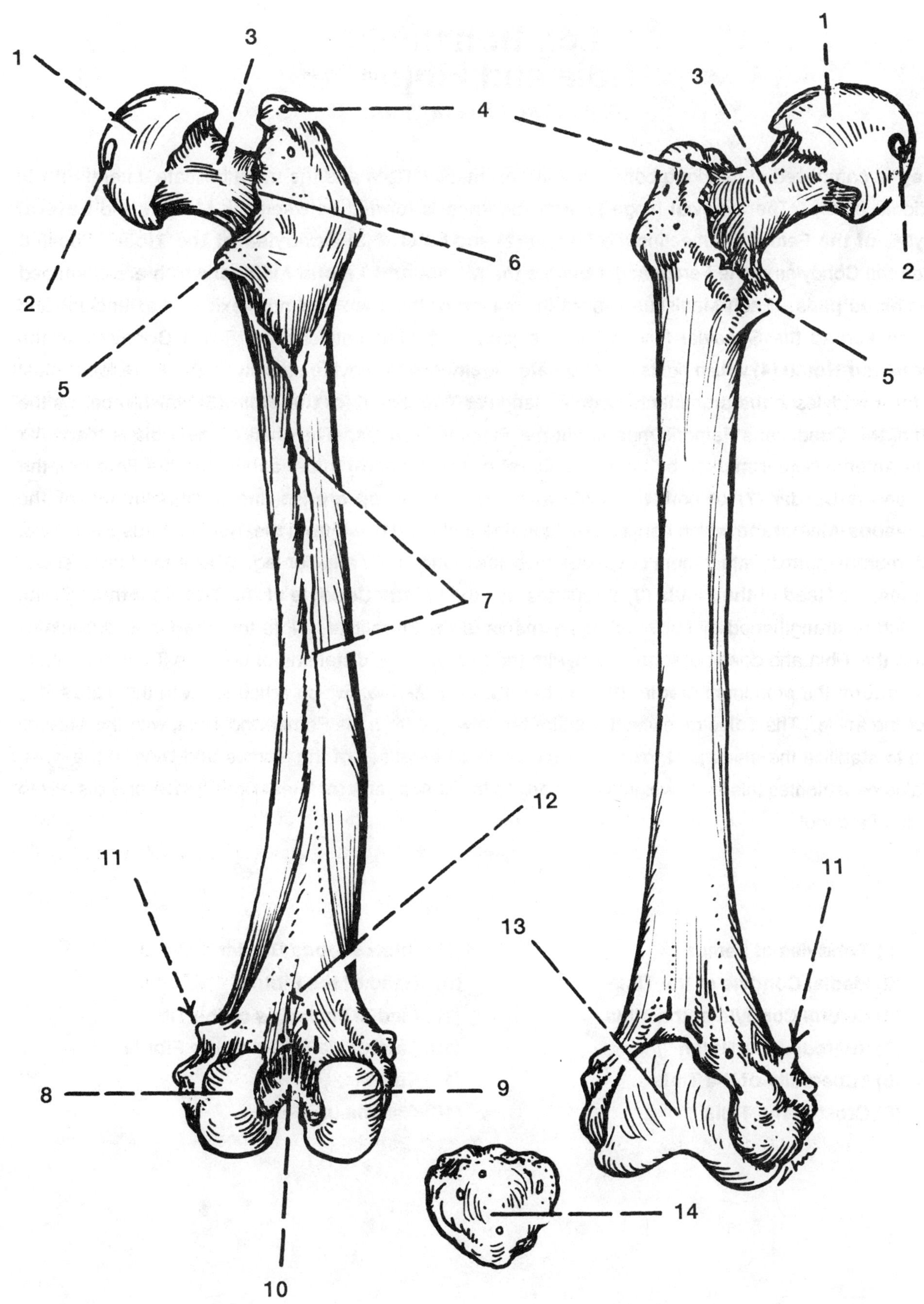

SS-14

Leg Bones:
Tibia and Fibula

(Right Side, Anterior View)

The **Leg** is composed of two long bones: the larger, medial **Tibia** and the more delicate, lateral **Fibula** ("the fiddle bow"). The synovial hinge joint of the knee is formed between the **Medial** and **Lateral Condyles of the Femur (1)** meeting the **Medial (2)** and **Lateral (3) Condyles of the Tibia**. Located between the Condyles of the Femur and Tibia are the **Medial and Lateral Menisci** which are C-shaped dense fibrous pads. Each Meniscus assists in rotation of the knee (but not flexion or extension) and serves to spread the Synovial fluid within the joint. Located between the Tibial Condyles is the **Intercondylar Notch (4)** which holds the **Cruciate Ligaments** to provide stability in the knee joint. Just below the Condyles is the prominent, roughly triangular **Tuberosity of the Tibia (5)** which receives the tendon of the Quadriceps Femoris muscle via the Patellar ligament. The shaft of the Tibia is triangular with the Anterior border marked by the sharp **Crest of the Tibia (6)**. Just as seen in the Forearm, the **Interosseous Border (7)** of both the Tibia and Fibula provides a crisp line of attachment for the **Interosseous Membrane** which contains oblique fibers sloping down from the Tibia towards the Fibula. This Membrane permits attachment of various muscles found in the anterior leg. Above the Interosseous Membrane, the **Head of the Fibula (8)** articulates with the **Lateral Condyle of the Tibia** to form a gliding joint which is strengthened by surrounding ligaments of the knee joint. Note that the Fibula articulates only with the Tibia and does not form a joint with the Femur. The distal end of both the Tibia and Fibula expand to form the prominent **Medial (9)** and **Lateral (10) Malleoli** which articulate with the **Talus (11)** bone of the ankle. The Talus receives the entire body weight from the Femur and Tibia, with the Malleoli serving to stabilize the ankle joint, much like the Styloid Processes of the Radius and Ulna at the wrist. The Talus redistributes this body weight posteriorly to the **Calcaneus** (or "Heel Bone") **(12)** and distally to the "ball" of the foot.

(1) Condyles of Femur
(2) Medial Condyle of the Tibia
(3) Lateral Condyle of the Tibia
(4) Intercondylar Notch
(5) Tuberosity of the Tibia
(6) Crest of the Tibia
(7) Interosseous Border
(8) Head of the Fibula
(9) Medial Malleolus of the Tibia
(10) Lateral Malleolus of the Fibula
(11) Talus
(12) Calcaneus

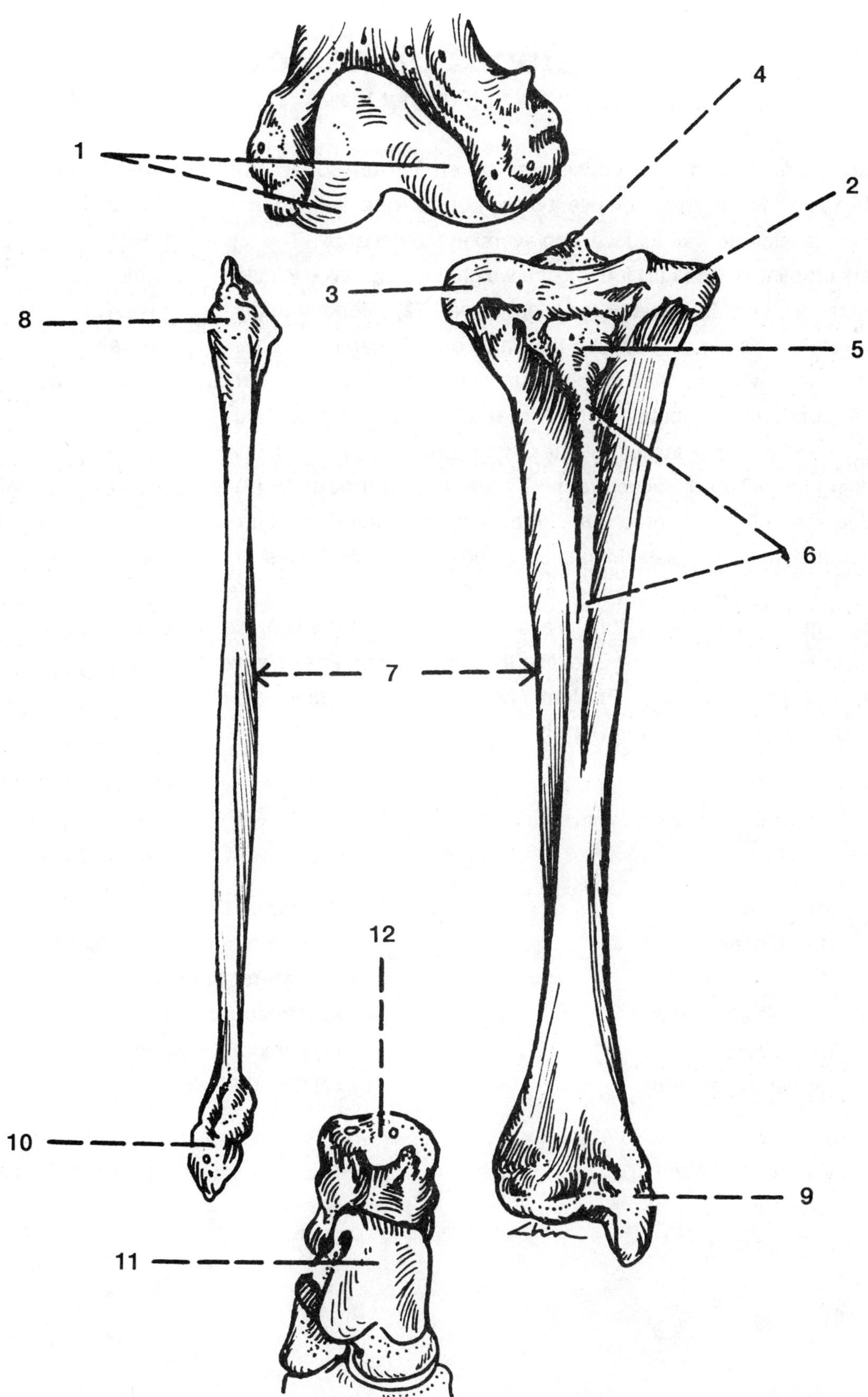
1
4
2
3
8
5
6
7
12
10
9
11

Bones of the Foot

(Superior View)

The skeleton of each Foot consists of seven **Tarsal** Bones, five **Metatarsal** Bones, and fourteen **Phalanges**, for a total of twenty six bones, one less than in the Hand. The Foot performs flexion/extension as well as inversion/eversion movements. The top of the Right Foot is pictured here, with the prominent **Talus (1)** forming the weight-bearing Ankle joint with the Tibia above. The Talus also articulates with the **Calcaneus** (or "Heel Bone") **(2)** inferiorly and with the **Navicular**(or Scaphoid) **(3)** Bone distally. The union between the Talus and Calcaneus is called the **Subtalar Joint (4)** because it occurs on the inferior surface of the Talus. The Calcaneus also articulates with the square **Cuboid (5)** bone in front and contributes to the **Midtarsal Joint (6)** of the Talus-Navicular and Calcaneus-Cuboid. The Calcaneus is the largest of the Tarsal Bones and is the only one to rest on the ground. The remaining three Tarsal Bones are the **Medial (7), Intermediate (8)** and **Lateral (9) Cuneiform Bones** ("wedge-shaped"). The three Cuneiforms along with the Cuboid articulate with the proximal ends of the Metatarsal bones. The distal Metatarsal Heads, especially the first and fifth, rest on the ground and form the "ball" of the foot. The five toes and the **Five Metatarsal Bones (10)** are numbered from medial to lateral. The First Metatarsal Bone is associated with the First Toe, also referred to as the "Great Toe" or Hallucis. Notice that the First Toe, like the Thumb, contains only two Phalangeal bones, the **Proximal Phalanx (11)** and the **Distal Phalanx (12)**, while each of the remaining toes include three Phalangeal bones (**Proximal, Middle, and Distal**).

(1) Talus
(2) Calcaneus
(3) Navicular
(4) Subtalar Joint
(5) Cuboid
(6) Midtarsal Joint
(7) Medial Cuneiform
(8) Intermediate Cuneiform
(9) Lateral Cuneiform
(10) Metatarsal Bones
(11) Proximal Phalanx
(12) Distal Phalanx

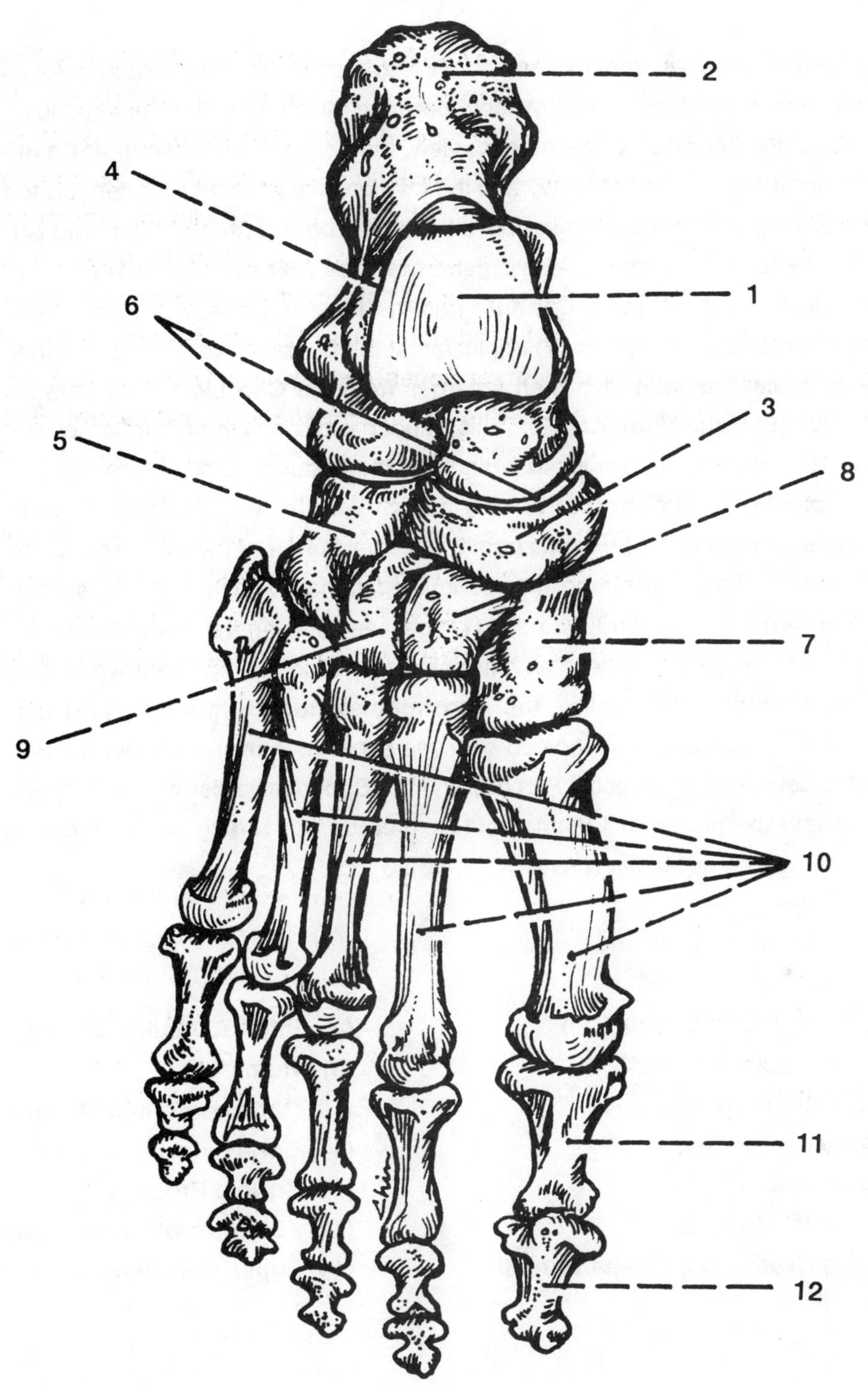
2
4
1
6
3
5
8
7
9
10
11
12

Shoulder Joint
(Anterior View)

The Right Shoulder is shown from the Anterior (Ventral) position. The Shoulder Joint is a Synovial, Ball-and-Socket type of Joint which is very unstable because the Head of the Humerus is much larger than the Glenoid Cavity of the Scapula. In this Anterior view, the belly of the **Subscapularis muscle (1)** has been cut to reveal the **Subscapular Fossa (2)** and the relationship of this muscle tendon to the Shoulder joint. Note that the Shoulder is protected from above by the **Acromion (3)** and **Coracoid (4) Processes of the Scapula**, the **Clavicle (5)**, and by various **Ligaments**. The **Articular Capsule (6) of the Shoulder Joint** consists of a dense outer **Fibrous Capsule** and the inner delicate **Synovial Membrane**. The **Long Head of the Biceps Brachii muscle (7)** originates at the Supraglenoid tubercle, with the Tendon passing through the **Articular Capsule**, and then emerges from the Capsule to lie in the **Intertubercular** (or Bicipital) **Groove (8)** of the **Humerus (9)**. This Tendon provides strong support over the anterior surface of the Head of the Humerus and helps to strengthen the Shoulder joint. Three Ligaments of the Shoulder link the **Clavicle (5)** to the **Acromion (3)** and **Coracoid (4) Processes** of the **Scapula**. The **Coracoacromial Ligament (10)** extends from the medial border of the Acromion to the lateral edge of the Corocoid Process. This is a strong flat triangular ligament which helps to support the Head of the Humerus. The **Coracoclavicular Ligament (11)** passes between the Corocoid Process and the **Conoid Process (12)** of the Clavicle. This is an extremely strong double ligament which helps to provide great stability to the shoulder joint. Lastly, the **Acromioclavicular Ligament (13)** extends between the Acromion and the Clavicle and is located over the synovial **Acromioclavicular Joint**. The **Scapular Notch**, located between the Corocoid Process and the Superior border of the Scapula, is bridged by the **Superior Transverse Scapular Ligament (14)**, thereby converting the Notch to a foramen for the transmission of the Suprascapular Nerve.

(1) Subscapularis Muscle (cut)
(2) Subscapular Fossa
(3) Acromion Process
(4) Coracoid Process
(5) Clavicle
(6) Articular Capsule
(7) Long Head of the Biceps Brachii

(8) Intertubercular Groove
(9) Humerus
(10) Coracoacromial Ligament
(11) Coracoclavicular Ligament
(12) Conoid Process
(13) Acromioclavicular Ligament
(14) Superior Transverse Ligament

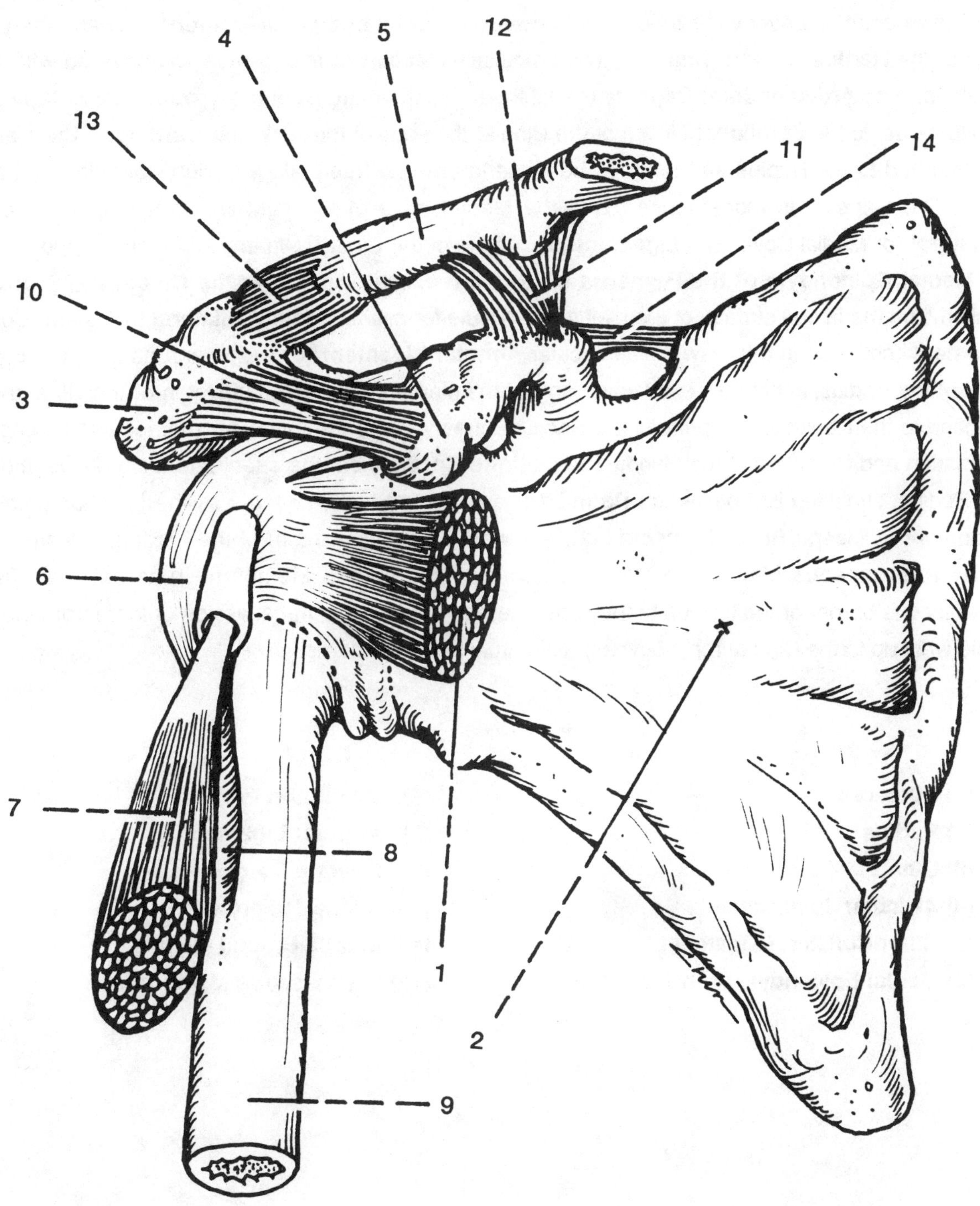

AS-1

Elbow Joint

(Left Elbow, Medial View)

The Elbow Joint is a synovial Hinge Joint between the distal end of the **Humerus (1)** and the proximal ends of the **Radius (2)** and **Ulna (3)**. The articular surfaces of these bones are covered with Hyaline Cartilage. The **Articular Joint Capsule (4)** extends from the margins of the **Trochlea** and **Capitulum** of the Humerus to the **Semilunar Notch** of the Ulna at the edge of the **Articular Cartilage.** The Capsule is not attached to the Radius and tends to be loose in order to freely allow flexion/extension of the elbow joint. The Capsule is more dense along the Medial edge of the joint where it is called the **Ulnar Collateral** (or Medial Collateral) **Ligament (5)**. Fibers of the Ulnar Collateral Ligament extend down from the **Medial Epicondyle of the Humerus (6)** to the proximal Ulna, near the **Olecranon Process (7)**. Similarly, on the lateral surface of the joint, the Capsule forms the **Radial Collateral** (or Lateral Collateral) **Ligament** (not seen in this view). The circular **Annular Ligament (8)** wraps around both the Head and Neck of the Radius, although it is attached only at the margins of the Radial Notch of the Ulna and is not attached to the Radius. This permits the Radius to freely rotate within the Annular Ligament in actions of Supination and Pronation of the Hand. The **Oblique Cord (9)** can be seen extending between the Ulna and Radius, attaching just below the **Radial Tuberosity (10)** which serves as the insertion point for the tendon of the **Biceps Brachii muscle (11)**, a powerful Flexor/Supinator of the Forearm. A small portion of the **Interosseous Membrane (12)** is pictured. This fibrous Membrane runs obliquely from the Interosseous border of the Radius to the Ulna, thereby allowing the transmission of force from the Radius at the wrist up to the Ulna at the elbow and ultimately to the Humerus.

(1) Humerus
(2) Radius
(3) Ulna
(4) Articular Joint Capsule
(5) Ulnar Collateral Ligament
(6) Medial Epicondyle of the Humerus
(7) Olecranon Process
(8) Annular Ligament
(9) Oblique Cord
(10) Radial Tuberosity
(11) Biceps Brachii muscle
(12) Interosseous Membrane

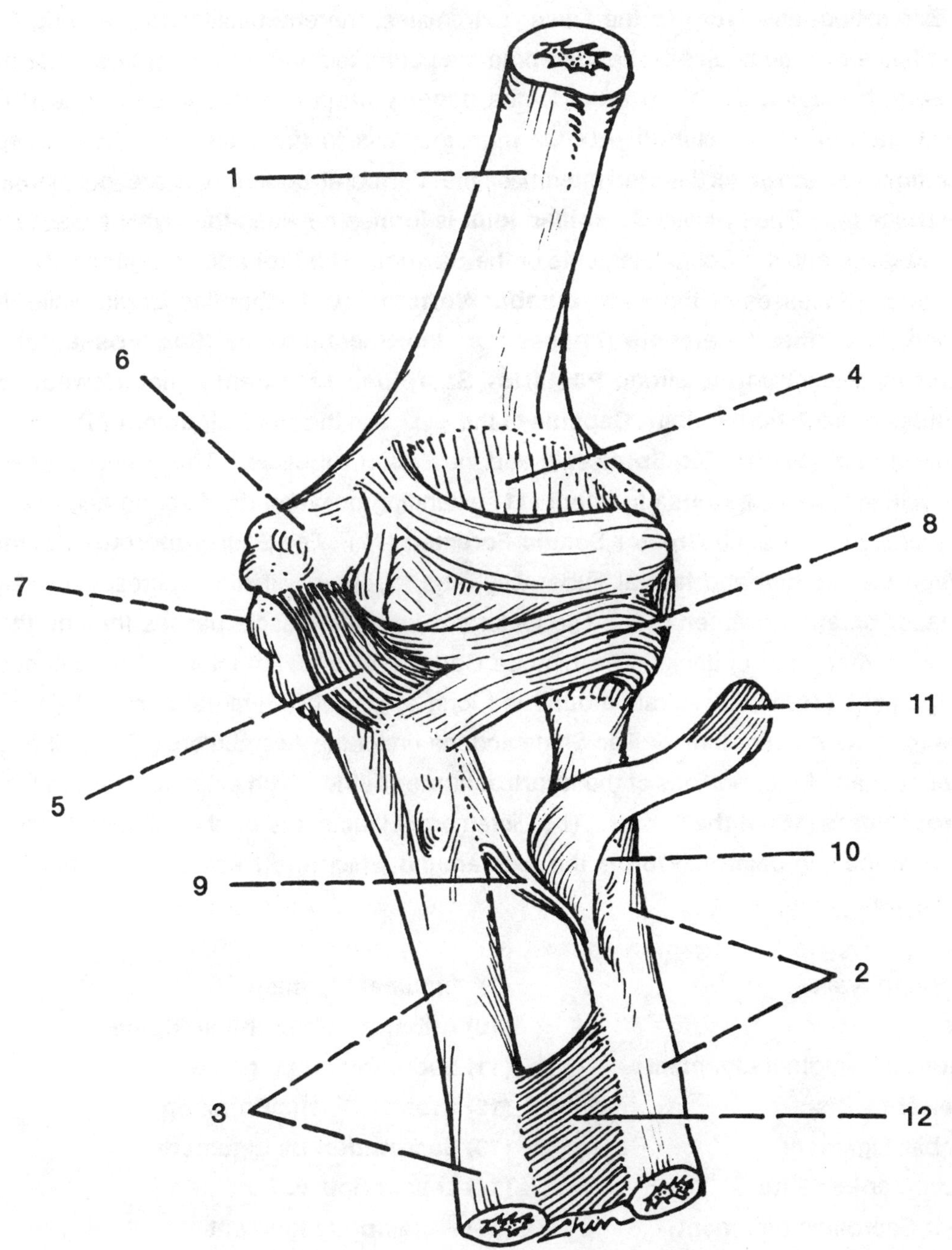
1
6
4
8
7
11
5
10
9
2
3
12
Chin

Pelvic Ligaments

(Anterior View)

The Pelvis if formed by the union of the right and left Hip bones joined anteriorly at the **Pubic Symphysis (1)** and posteriorly the Hip bones are separated by the Sacrum and Coccyx of the Vertebral Column. Inferiorly, the Pelvis articulates with the **Femurs (2)** which permits the transfer of weight from the Head, Neck, Upper Extremities, and Trunk to the Lower Extremities, thereby facilitating the upright posture characteristic of humans. The bodies of the Vertebrae are bound together by the **Anterior Longitudinal Ligament (3)** which extends from the Atlas to the upper portion of the Sacrum. The **Posterior Longitudinal Ligament** (not pictured) extends from the Axis to the Sacrum. These Longitudinal Ligaments are firmly attached to the Periosteum of the Vertebral bodies, yet are not bound to the **Intervertebral Discs (4)**. The synovial **Sacroiliac joint** is formed between the Hyaline Cartilage of the Ilium Auricular surfaces and the Costal elements of the Sacrum. The **Iliolumbar Ligaments (5)** extend from the transverse processes of the **Fifth Lumbar Vertebra (6)** to the Iliac Crest, while the more flattened **Anterior Sacroiliac Ligaments (7)** pass from the Sacrum to the **Iliac Fossa (8)**. These Anterior Ligaments, along with the strong **Posterior Sacroiliac Ligaments** (not viewed), serve to surround and support the **Articular Joint Capsule** of the Hip. The **Inguinal Ligament (9)** runs obliquely between the **Anterior Superior Iliac Spine (10)** and the **Pubic Tubercle**. The Spine of the Ischium points medially with the **Sacrospinous Ligament (11)** extending between the Sacrum and the tip of the Ischial Spine, thereby creating the **Greater Sciatic Foramen (12)**. The **Sacrotuberous Ligament (13)** extends between the Sacrum and Ischial Tuberosity, and together with the Sacrospinous Ligament, creates the **Lesser Sciatic Foramen (14)**. The Obturator Internus muscle passes through the Lesser Sciatic Foramen to reach the buttock. The Fibrous Capsule of the Hip Joint is strengthened by the **Iliofemoral Ligament (15)** which spirals around the long axis of the Femoral neck. This Y-shaped ligament arises from the Anterior Inferior Iliac Spine and the rim of the Acetabulum. The split ends of the Y attach to the upper and lower portions of the **Intertrochanteric line** which extends between the Greater and **Lesser Trochanters (16) of the Femur**. The Iliofemoral Ligament is of clinical importance in that it limits extension of the Hip Joint and forms the axis around which the Neck of the Femur rotates in dislocation of this joint.

(1) Pubic Symphysis
(2) Femur
(3) Anterior Longitudinal Ligament
(4) Intervertebral Disc
(5) Iliolumbar Ligament
(6) Fifth Lumbar Vertebra
(7) Anterior Sacroiliac Ligament
(8) Iliac Fossa
(9) Inguinal Ligament
(10) Anterior Superior Iliac Spine
(11) Sacrospinous Ligament
(12) Greater Sciatic Foramen
(13) Sacrotuberous Ligament
(14) Lesser Sciatic Foramen
(15) Iliofemoral Ligament
(16) Lesser Trochanter

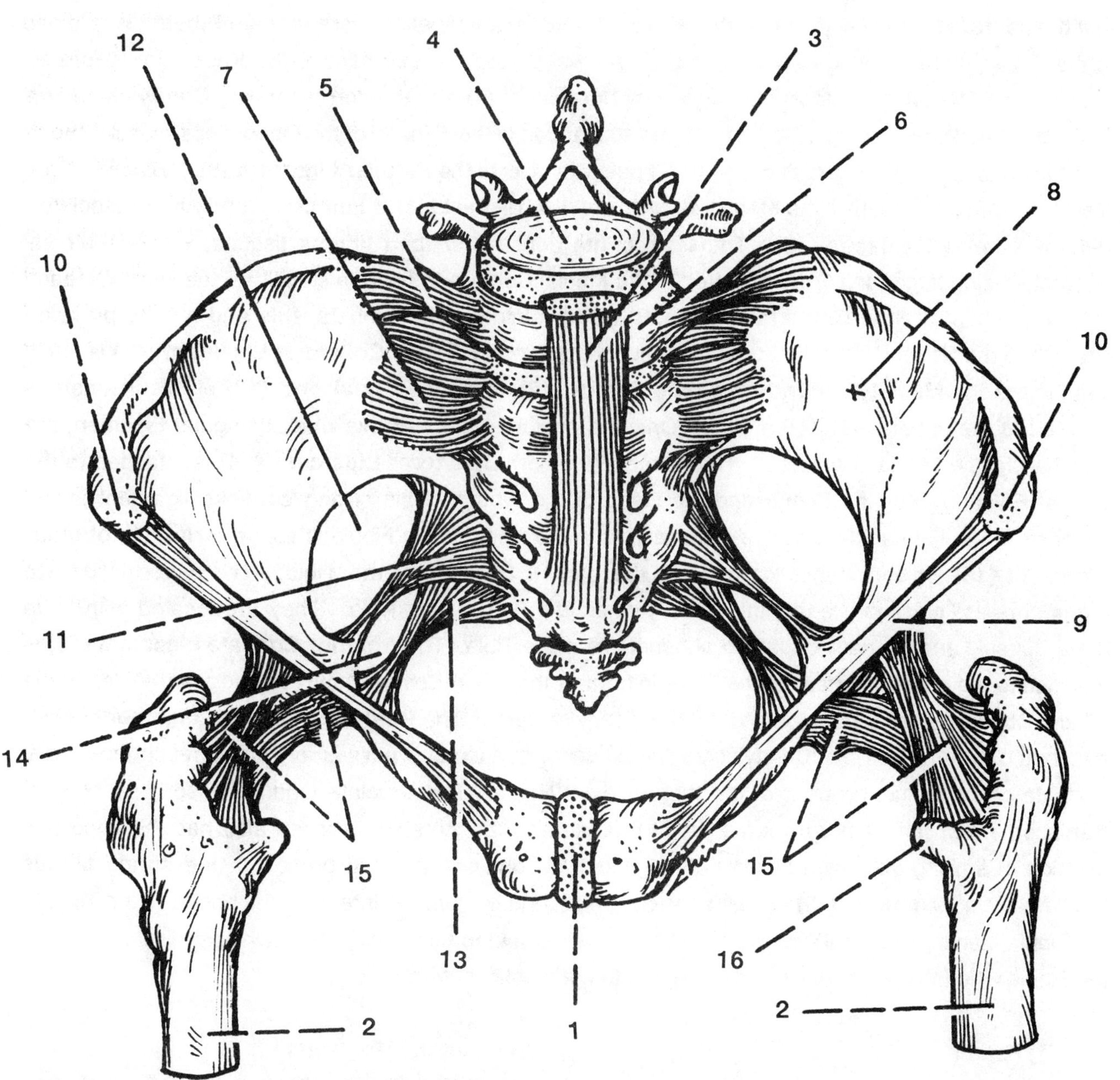
12
7
5
4
3
6
8
10
10
11
9
14
15
15
13
16
2
1
2

Knee Joint

(Anterior View)

In this view of the Left Knee Joint, only the distal end of the Femur is shown and the Tendon of the Quadriceps Muscle (suspending the Patella) has been cut to reveal the contents of the Joint cavity. Note that the **Femur (1)** articulates only with the **Tibia (2)**, while the smaller, more delicate **Fibula (3)** is joined to the Tibia (not articulating with the Femur). The weight-bearing segment of the Knee joint is formed between the **Medial and Lateral Condyles of the Femur (4)** and the corresponding **Condyles of the Tibia (5)**. The **Patella (6)** is a Sesamoid bone suspended in the Tendon of the Quadriceps muscle above and it is attached to the **Tibial Tuberosity (7)** below by way of the **Patellar Ligament (8)**. Adjacent to the **Medial Femoral Condyle** is the **Medial Meniscus (9)**, while the **Lateral Femoral Condyle** is associated with the **Lateral Meniscus (10)**. The Menisci are dense C-shaped fibrous tissues, with the Medial Meniscus being larger and firmly attached to the Medial Condyle to provide stability to the joint. On each side of the joint are the Collateral Ligaments which reinforce the Joint Capsule and help to provided strength to the Knee Joint. The **Proximal end** of the **Medial** (Tibial) **Collateral Ligament (11)** is firmly attached to the **Medial Epicondyle of the Femur (12)**, while the **Distal end (13)** of this ligament is secured to the upper shaft of the Tibia. Only the posterior fibers of this wide, triangular Ligament are attached to the Medial Meniscus. The **Lateral** (Fibular) **Collateral Ligament (14)** is attached to the Lateral Epicondyle of the Femur and the Head of the Fibula. This strong cord-like ligament is not attached to the Lateral Meniscus. Also attached to the Head of the Fibula is the **Anterior Tibiofibular Ligament (15)** which extends down from the Lateral Condyle of the Tibia. The paired **Cruciate Ligaments** are crossed like the letter X to connect the Femur and Tibia. They are located within the Knee Capsule and are named for their attachments to the Tibia. The **Anterior Cruciate Ligament (16)** is attached to the anterior surface of the Tibia, in front of the Tibial Spine. It curves backward to reach the Lateral Condyle of the Femur, behind the Intercondylar Notch. The **Posterior Cruciate Ligament (17)** is attached between the Tibial Condyles on the posterior surface and passes forward to reach the Medial Condyle of the Femur, in front of the Intercondylar Notch. The Cruciate Ligaments serve to prevent displacement of the Femur from the Tibial Plateau and the Anterior Cruciate also has the important function of limiting extension of the Lateral Femoral Condyle. A small portion of the strong fibrous **Interosseous Membrane (18)** can be seen running obliquely from the Interosseous border of the Tibia to the Fibula, serving to bind these two leg bones together and to help resist the downward displacement of the Fibula when the powerful muscles attached to the Fibula contract.

(1) Femur
(2) Tibia
(3) Fibula
(4) Medial and Lateral Condyles of the Femur
(5) Condyle of the Tibia
(6) Patella
(7) Tibial Tuberosity
(8) Patellar Ligament
(9) Medial Meniscus
(10) Lateral Meniscus
(11) Medial Collateral Ligament (Proximal end)
(12) Medial Epicondyle of the Femur
(13) Medial Collateral Ligament (Distal end)
(14) Lateral Collateral Ligament
(15) Anterior Tibiofibular Ligament
(16) Anterior Cruciate Ligament
(17) Posterior Cruciate Ligament
(18) Interosseous Membrane

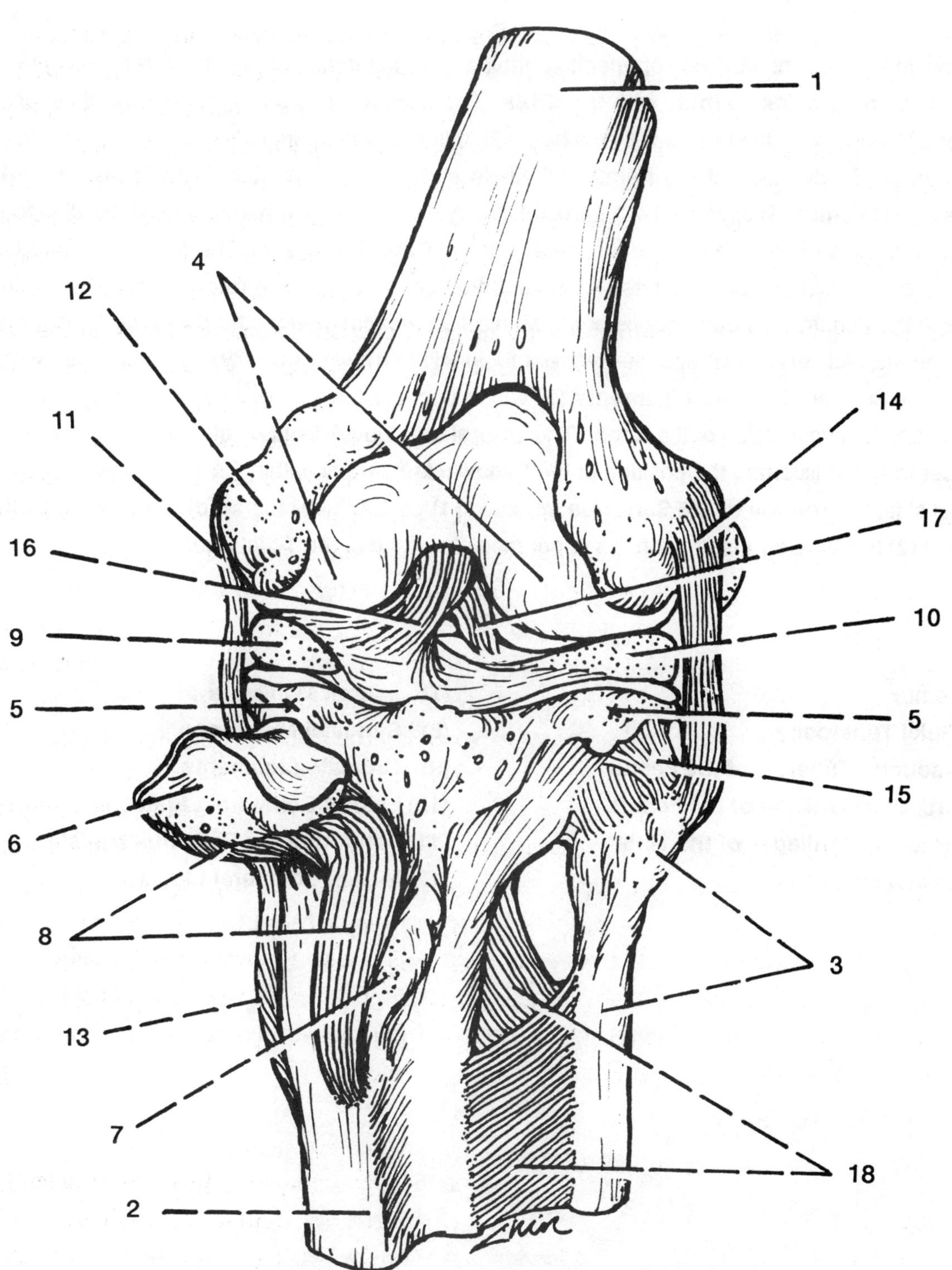

AS-4

Knee Joint

(Interior View)

In this view, the Knee joint has been opened to permit an understanding of the structures uniting the distal **Femur (1)** to the proximal **Tibia**, with the **Tibial Tuberosity (2)** serving as a reference point for orientation. Various bony landmarks will also help with this orientation, including the **Adductor Tubercle (3)** found on the Medial side of the Femur. A single plate of **Hyaline Cartilage** forms the **Articular Cartilage of the Femur (4)** covering both Femoral Condyles, while two separate caps of Hyaline form the **Articular Cartilages of the Tibia (5)** covering each of the **Tibial Condyles**. The **Medial (6)** and **Lateral Menisci (7)** are positioned between these Articular Cartilages and assist in rotation of the Knee Joint by moving with the Femur. The outer edges of the Menisci are not covered by the **Synovial Membrane (8)**, although the Menisci serve to spread the Synovial Fluid within the Joint Cavity. Located deep within the Cavity are the crossed **Cruciate Ligaments (9)** which connect the Femur and Tibia, acting to prevent dislocation and hyperextension of the Knee. The **Lateral** (or Fibular) **Collateral Ligament (10)** has been cut in order to better illustrate the position of the Synovial Membrane within the Fibrous Joint Cavity. On the Medial side, the **Tendon of the Sartorius muscle (11)** passes over the **Medial** (or Tibial) **Collateral Ligament (12)** to ultimately insert upon the upper medial surface of the Tibial Shaft.

(1) Femur
(2) Tibial Tuberosity
(3) Adductor Tubercle of the Femur
(4) Articular Cartilage of the Femur
(5) Articular Cartilages of the Tibia
(6) Medial Meniscus
(7) Lateral Meniscus
(8) Synovial Membrane
(9) Cruciate Ligaments
(10) Lateral Collateral Ligament
(11) Tendon of the Sartorius Muscle
(12) Medial Collateral Ligament

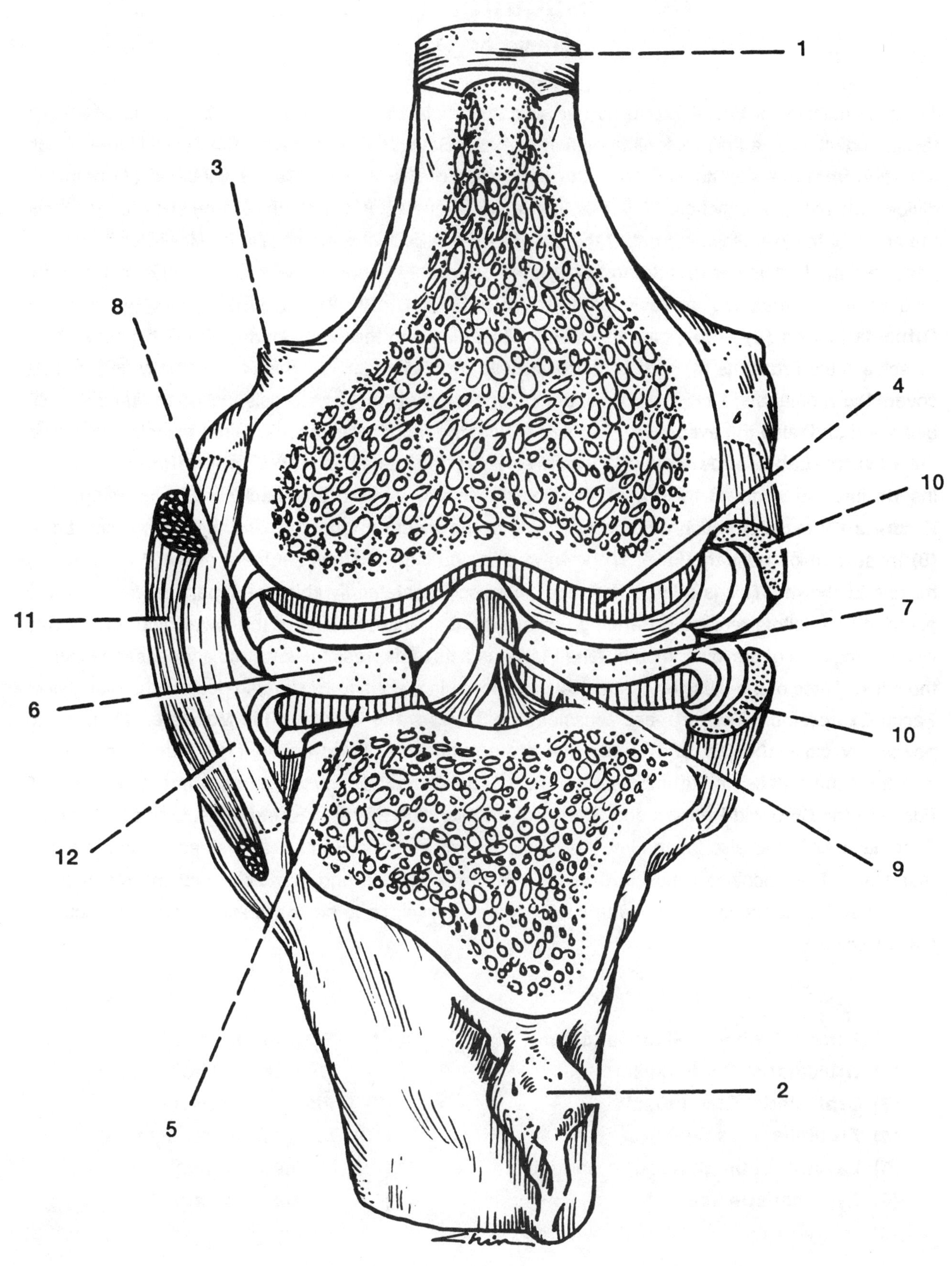

AS-5

Muscles of Facial Expression and Mastication (chewing)

(Anterior View)

All of the muscles of Facial Expression are thin superficial muscles located within the subcutaneous fascia and all receive their motor innervation from the Seventh Cranial nerve, the Facial Nerve, which emerges from the skull to penetrate the substance of the **Parotid Salivary Gland (1)** before it divides into multiple branches. Note that the Parotid Gland extends from the Zygomatic Arch to below the angle of the Mandible, in front of the ear. The Parotid Duct runs across the Masseter muscle to penetrate the Buccinator muscle and enters the oral cavity opposite the second upper molar tooth. Two Sphincter muscles are evident, the **Orbicularis Oculi (2)** surrounding the Eye and the **Orbicularis Oris (3)** surrounding the Mouth. Inserting into the upper part of the Orbicularis Oculi muscles is the **Frontalis (4)** portion of the Occipito-Frontalis muscle. This is a two-belly muscle that covers the Frontal and Occipital bones, being connected by an Aponeurosis, acting to pull the scalp backward and raise the eyebrows. Associated with the mouth are various paired muscles, including the **Levator Labii Superioris (5)** which elevates the upper lip, the **Zygomaticus (6)** which inserts into the angle of the mouth drawing the mouth upward and backward in laughing, the **Risorius (7)** muscles which retracts the angle of the mouth in grinning, and the **Depressor Labii (8)** muscle which pulls the lower lip downward in pouting! A small segment of the **Platysma (9)** muscle is shown. This is a large flat sheet of muscle contained within the superficial fascia of the pectoral and deltoid regions, extending upward to insert into the skin of the lower face and blends with the muscles of the lower mouth. The Platysma acts to depress the lower lip and tense the skin of the neck. Three of the Muscles of Mastication (or chewing) are shown situated deep to the muscles of Facial Expression. Two of these muscles, the **Temporalis (10)** and the **Masseter (11)**, act to powerfully close the jaw and clench the teeth, while the **Buccinator (12)** muscle serves to compress the cheeks (the "trumpeters muscle"). The Temporalis muscle extends from the Temporal Fossa to the Corocoid Process and Ramus of the Mandible. The Masseter muscle arises from the Zygomatic Arch and also inserts into the Corocoid Process and Ramus, as well as the angle of the Mandible. The Buccinator muscle originates from the Maxilla and Mandible near the Molar tooth sockets and its fibers converge towards the angle of the mouth to blend with fibers of the Orbicularis Oris muscle.

(1) Parotid Salivary Gland and Duct
(2) Orbicularis Oculi muscle
(3) Orbicularis Oris muscle
(4) Frontalis muscle
(5) Levator Labii Superioris muscle
(6) Zygomaticus muscle
(7) Risorius muscle
(8) Depressor Labii muscle
(9) Platysma muscle
(10) Temporalis muscle
(11) Masseter muscle
(12) Buccinator muscle

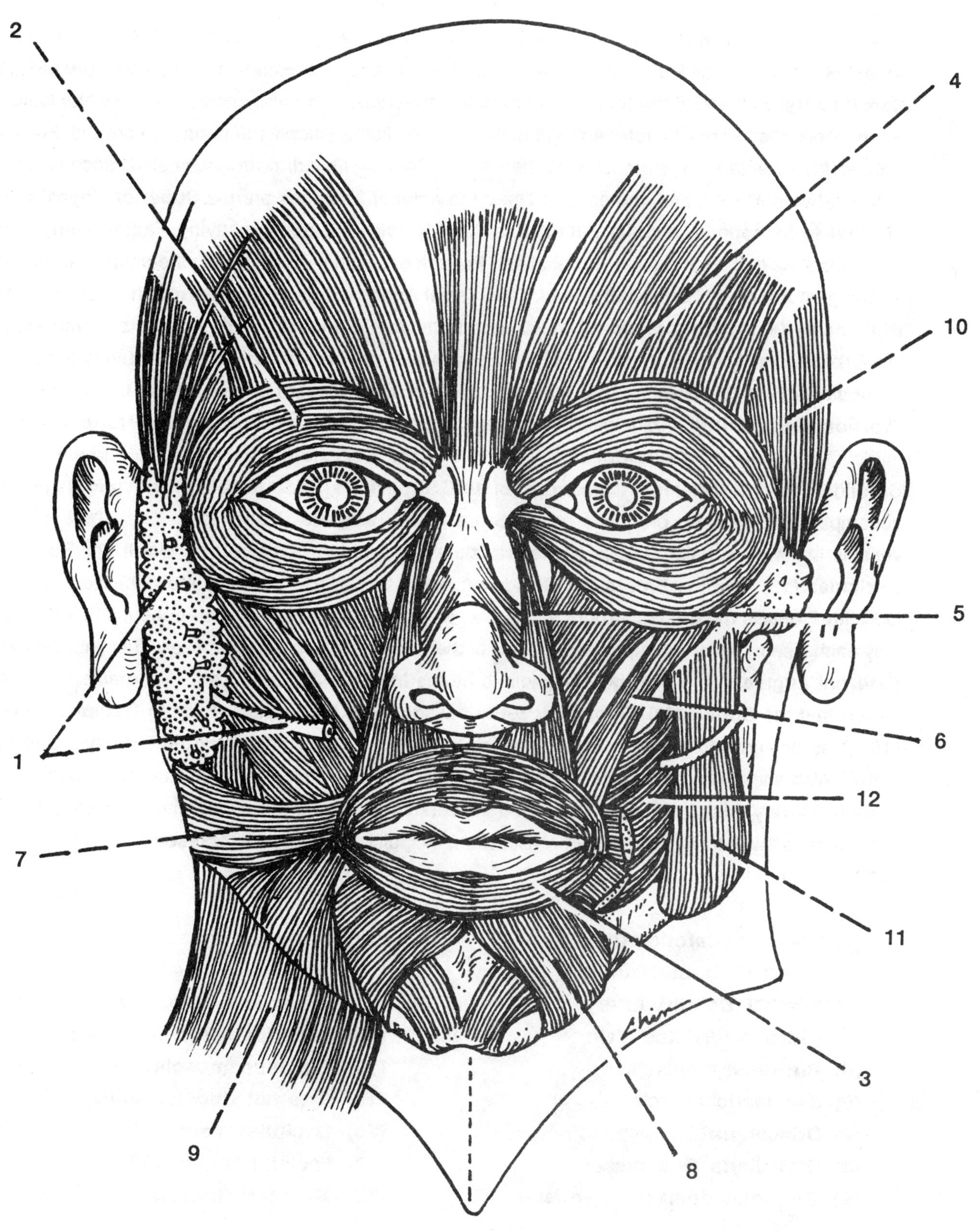

MS-1

Superficial Muscles of the Face and Neck

(Lateral View)

The entire head and neck is provided with arterial blood by way of the bilateral **Common Carotid Arteries (1)** which divide near the angle of the jaw into the **External (2)** and **Internal (3) Carotid Arteries**. The Superficial Facial Muscles are supplied by branches of the External Carotid Artery, while the Internal Carotid Arteries penetrate the skull to supply the brain. Before the External Carotid Artery enters the substance of the Parotid Salivary Gland, it produces six branches, three anterior, two posterior, and one deep. The three anterior branches are the **Superior Thyroid**, the **Lingual** to the tongue, and the **Facial Arteries (4)** (with the accompanying **Facial Vein**). The three posterior branches are the Occipital and Posterior Auricular Arteries, with the deeper Ascending Pharyngeal Artery which are not visible. The Facial Artery spirals upward across the face, over the **Buccinator (5)** and **Zygomatic (6)** muscles to terminate at the inner corner of the eye and supply the **Orbicularis Oculi (7)** muscle . The Lingual Artery crosses the lower face to terminate at the **Orbicularis Oris (8)** muscle. The Occipital Artery courses upward to give off two branches to the **Sternocleidomastoid (9)** muscle and then continues upward, medial to the **Trapezius (10)** muscle, to supply the **Occipitalis (11)** portion of the Occipito-Frontalis muscle. The External Carotid Artery terminates in branches of the **Superficial Temporal Artery (12)** which is accompanied by similar branches of the **Superficial Temporal Vein (12)**. Branches of these vessels are seen running across the lateral skull to supply the **Orbicularis Oculi (7), Superior Auricular (13)**, and the **Frontalis (14)** portion of the **Occipito-Frontalis** muscle. Note that the Occipito-Frontalis muscle consists of two bellies which are separated by an Aponeurosis into which they each insert. All of the superficial vessels of the Face and Neck will collectively drain into the large **External Jugular Vein (15)** which is joined by the **Internal Jugular Vein** that drains the brain. Notice that the **Superficial Temporal Veins** pass anterior to the ear, while the **Occipital Veins (16)** pass posterior to the ear. These multiple branches join to help form the **External Jugular Vein**. Also shown in this region is the **Facial Nerve (17)** passing through the substance of the Parotid Salivary Gland and then branching repeatedly to supply the superficial muscles of Facial Expression as well as the Muscles of Mastication, including the powerful **Masseter (18)** muscle.

(1) Common Carotid Artery
(2) External Carotid Artery
(3) Internal Carotid Artery
(4) Facial Artery and Vein
(5) Buccinator muscle
(6) Zygomatic muscle
(7) Orbicularis Oculi muscle
(8) Orbicularis Oris muscle
(9) Sternocleidomastoid muscle
(10) Trapezius muscle
(11) Occipitalis muscle
(12) Superficial Temporal Vessels
(13) Superior Auricular muscle
(14) Frontalis muscle
(15) External Jugular Vein
(16) Occipital Vein
(17) Facial Nerve
(18) Masseter muscle

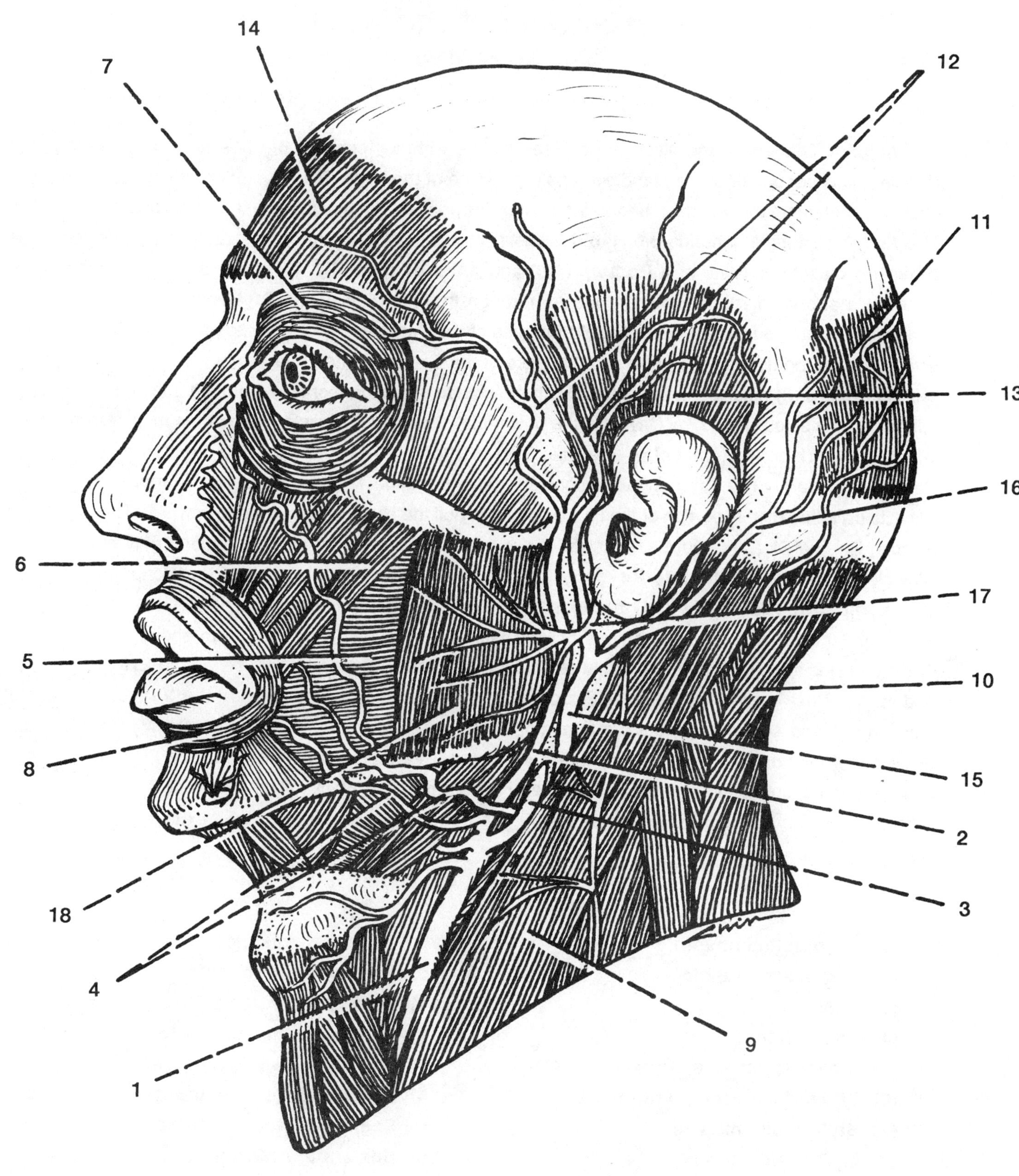

MS-2

Anterior and Posterior Triangles of the Neck

(Lateral View)

The Anterior Triangle of the Neck is bordered by the anterior edge of the **Sternocleidomastoid (1)** muscle, the lower border of the Mandible, and the midline of the Neck. The Posterior Triangle of the Neck is enclosed between the Sternocleidomastoid and **Trapezius (2)** muscles, with the **Clavicle (3)** below. The Anterior Triangle contains the **Hyoid Bone (4)** for the attachment of various tongue muscles. The **Digastric** muscle consists of a **Posterior (5)** and **Anterior (6) belly** which are anchored to the Hyoid bone by a strong Intertendon which passes through the insertion of the **Stylohyoid (7)** muscle. The Stylohyoid muscle originates at the Styloid Process of the Temporal Bone and, together with the Digastric, acts to elevate both the Hyoid bone and the Base of the Tongue. The Floor of the Mouth is formed by the two **Mylohyoid (8)** muscles which extend from the chin to the Hyoid bone. Extending above the Mandible, reinforcing the sides of the Oral Cavity, is the powerful **Masseter (9)** muscle. Two of the muscles that insert into the Hyoid bone from below are the **Sternohyoid (10)** muscles and the **Omohyoid (11)** ("omo"=shoulder). The Sternohyoid muscles extend from the Manubrium of the Sternum and medial Clavicle up to the medial portion of the Hyoid, acting to pull the Hyoid and Larynx downward. The Omohyoid muscle, like the Digastric muscle, consists of two bellies separated by an Intertendon firmly bound to the Clavicle. The **Omohyoid Inferior belly (12)** begins at the upper border of the Scapula, while the **Omohyoid Superior belly (11)** ascends to insert into the lower border of the Hyoid bone. In the Posterior Triangle of the Neck, portions of four muscles are shown. Starting from above, the **Splenius Capitis (13)** is seen passing up to the Mastoid Process of the Temporal bone, serving to Extend the Head and Neck. Next is seen the **Levator Scapulae (14)** descending to insert upon the Vertebral border of the Scapula and acting to elevate the Scapula, as the name indicates. The **Scalenus Medius (15)** and **Scalenus Anterior (16)** muscles originate from Cervical Vertebrae and insert into the First rib, acting to elevate this rib and assists in both Flexion and Rotation of the Neck. The Posterior Triangle contains numerous Lymph Nodes, the Subclavian Artery, the three trunks of the Brachial Plexus, and Loops of the Cervical Plexus.

(1) Sternocleidomastoid muscle
(2) Trapezius muscle
(3) Clavicle
(4) Hyoid bone
(5) Digastric muscle, Posterior belly
(6) Digastric muscle, Anterior belly
(7) Stylohyoid muscle
(8) Mylohyoid muscle
(9) Masseter muscle
(10) Sternohyoid muscle
(11) Omohyoid muscle, Superior belly
(12) Omohyoid muscle, Inferior belly
(13) Splenius Capitis muscle
(14) Levator Scapulae muscle
(15) Scalenus Medius muscle
(16) Scalenus Anterior muscle

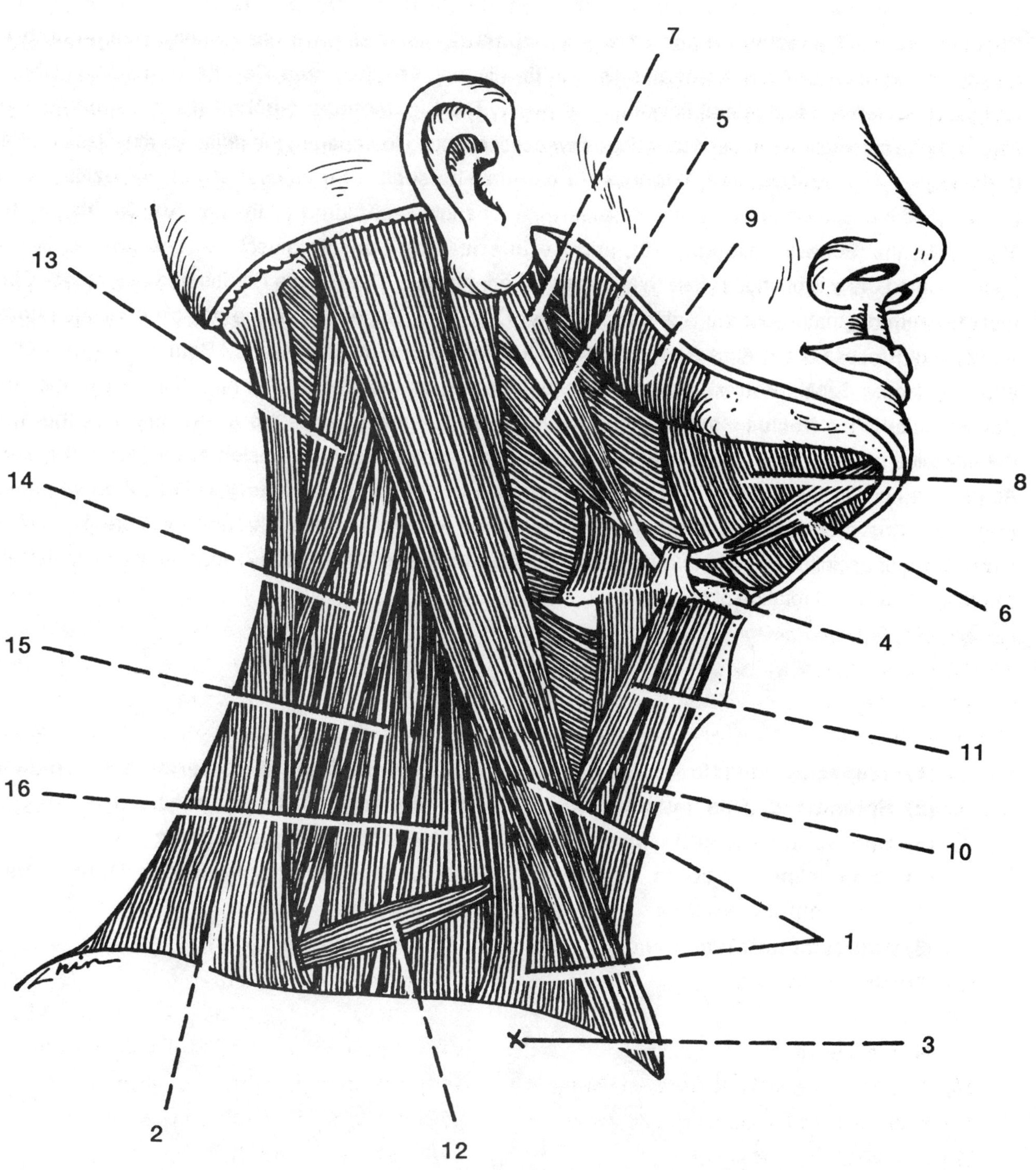

MS-3

Superficial and Deep Posterior Neck Muscles

The Superficial muscles are pictured on the Left side, with the Deeper muscles seen on the Right. Segments of the **Trapezius (1)** and **Splenius Capitis (2)** muscles have been removed in order to more clearly display the powerful **Semispinalis Capitis (3)** and **Longissimus Capitis (4)** muscles which act together in extension and lateral flexion of the neck. The **Ligamentum Nuchae (5)** is a strong triangular fibrous mass connecting the Cervical spinous processes, acting to separate the adjacent muscles and Fascia of the neck. The **Suboccipital Triangle (6)** can only be seen with removal of the Trapezius, Splenius Capitis, and the Semispinalis Capitis muscles. The Triangle is bordered by the **Rectus Capitis Posterior Major (7)**, the **Superior Oblique (8)**, and the **Inferior Oblique (9)** muscles. The Rectus Capitis is poorly named, as it is oblique rather than vertical in orientation. The Rectus Capitis and the Inferior Oblique muscles both originate from the Spine of the Axis (C-2). The Superior Oblique muscle originates from the Transverse Process of the Atlas (C-1), where the Inferior Oblique muscle inserts. Both the Superior Oblique and the Rectus Capitis Posterior Major muscles insertion on the Occipital bone. Adjacent to the Rectus Capitis Major is the **Rectus Capitis Posterior Minor (10)** which extends more vertically from the Atlas to the Occipital bone. These muscles of the Suboccipital Triangle act in extension and rotation of the Head. Across the floor of the Triangle can be seen a portion of the **Vertebral Artery (11)** as it curves medially from the Transverse foramina of the Cervical Vertebrae to enter the skull through the Foramen Magnum. Below the Suboccipital Triangle is a portion of the **Semispinalis Cervicis (12)** muscle which assists in the extension and lateral rotation of the Head.

(1) Trapezius muscle
(2) Splenius Capitis muscle
(3) Semispinalis Capitis muscle
(4) Longissimus Capitis muscle
(5) Ligamentum Nuchae
(6) Suboccipital Triangle
(7) Rectus Capitis Posterior Major muscle
(8) Superior Oblique muscle
(9) Inferior Oblique muscle
(10) Rectus Capitis Posterior Minor muscle
(11) Vertebral Artery
(12) Semispinalis Cervicis muscle

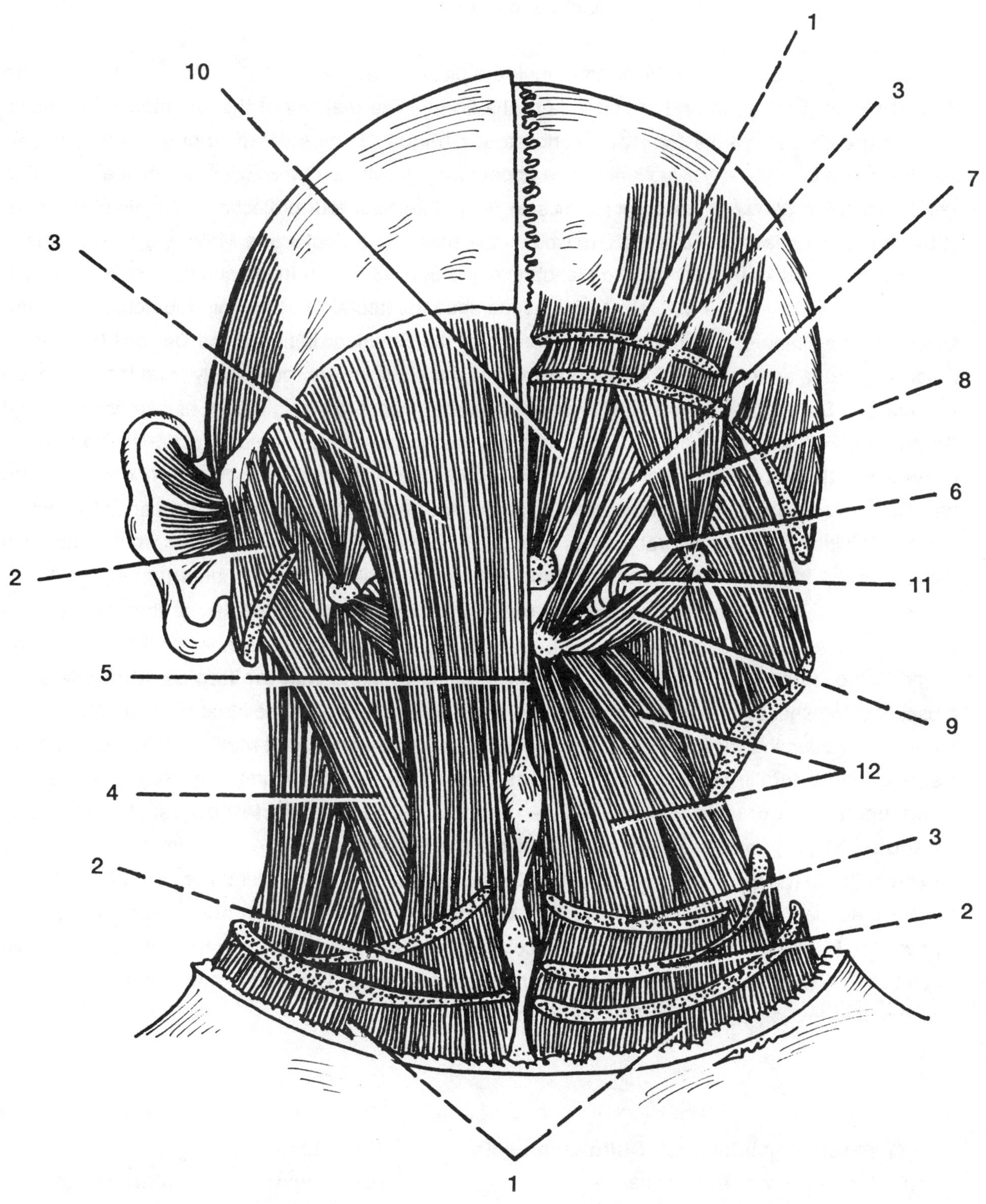

MS-4

Anterior Muscles of the Trunk

(Superficial and Deep Muscles)

The Superficial Muscles of the Neck and Trunk are pictured on the Right side of the body, while the Deeper Muscles are seen on the Left side of the body. The majority of the superficial Abdominal musculature consists of the **External Abdominal Oblique (1)** muscle which originates as finger-like digitations from the lower eight ribs. The upper four projections of the External Oblique intertwine with fibers of the **Serratus Anterior (2)** muscle, while the lower four projections are joined by fibers of the massive **Latissimus Dorsi (3)** muscle of the back. The **Pectoralis Major (4)** muscle arises from the medial half of the clavicle, the Manubrium and Body of the Sternum, and the Costal cartilages of the upper six ribs. The Pectoralis Major inserts upon the lateral lip of the Intertubercular (Bicipital) Groove of the Humerus. Forming the roundness of the Shoulder is the dense **Deltoid (5)** muscle which originates anteriorly from the lateral third of the clavicle, and posteriorly from the Acromion Process and Spine of the Scapula. This massive muscle tapers to a single point of insertion upon the Deltoid Tuberosity of the Humerus. The Deltoid is said to have three separate actions: The anterior fibers assist the Pectoralis Major muscle in Flexion of the Shoulder, the posterior fibers assist the Latissimus Dorsi in Extension of the Arm, while the Acromial fibers Abduct the Arm once the action has been initiated by the Supraspinatus muscle. Immediately deep to the External Abdominal Oblique is the **Internal Abdominal Oblique (6)** muscle which contains fibers running at nearly 90° to the External. The **Transverse Abdominis (7)** muscle is the deepest of these abdominal wall muscles, running horizontally from the Costal Cartilages of the lowest six ribs and Iliac Crest to the midline **Linea Alba (8)**. These three Abdominal muscles along with the **Rectus Abdominis (9)** muscle act together to compress the Abdomen and help support the Abdominal Viscera. Notice that the Rectus Abdominis includes three Tendinous Intersections which end medially at the Linea Alba. Deep to the Pectoralis Major is the **Pectoralis Minor (10)** muscle arising from Ribs 3, 4 and 5 to insert upon the Coracoid Process of the Scapula. The **Sternocleidomastoid (11)** and **Sternohyoid (12)** muscles both originate from the Manubrium of the Sternum and the medial portion of the Clavicle. The upper end of the Deltoid muscle is said to extend above the Clavicle into the Neck as the **Trapezius (13)** muscle. The **Biceps Brachii (14)** muscle arises from the Scapula by two Heads: the Short Head from the Corocoid Process, and the Long Head from the Supraglenoid Tuberosity. The Biceps Brachii muscle inserts upon the Radial Tuberosity of the Humerus thereby acting across two joints, the Shoulder and Elbow.

(1) External Abdominal Oblique muscle
(2) Serratus Anterior muscle
(3) Latissimus Dorsi muscle
(4) Pectoralis Major muscle
(5) Deltoid muscle
(6) Internal Abdominal Oblique muscle
(7) Transverse Abdominis muscle
(8) Linea Alba
(9) Rectus Abdominis muscle
(10) Pectoralis Minor muscle
(11) Sternocleidomastoid muscle
(12) Sternohyoid muscle
(13) Trapezius muscle
(14) Biceps Brachii muscle

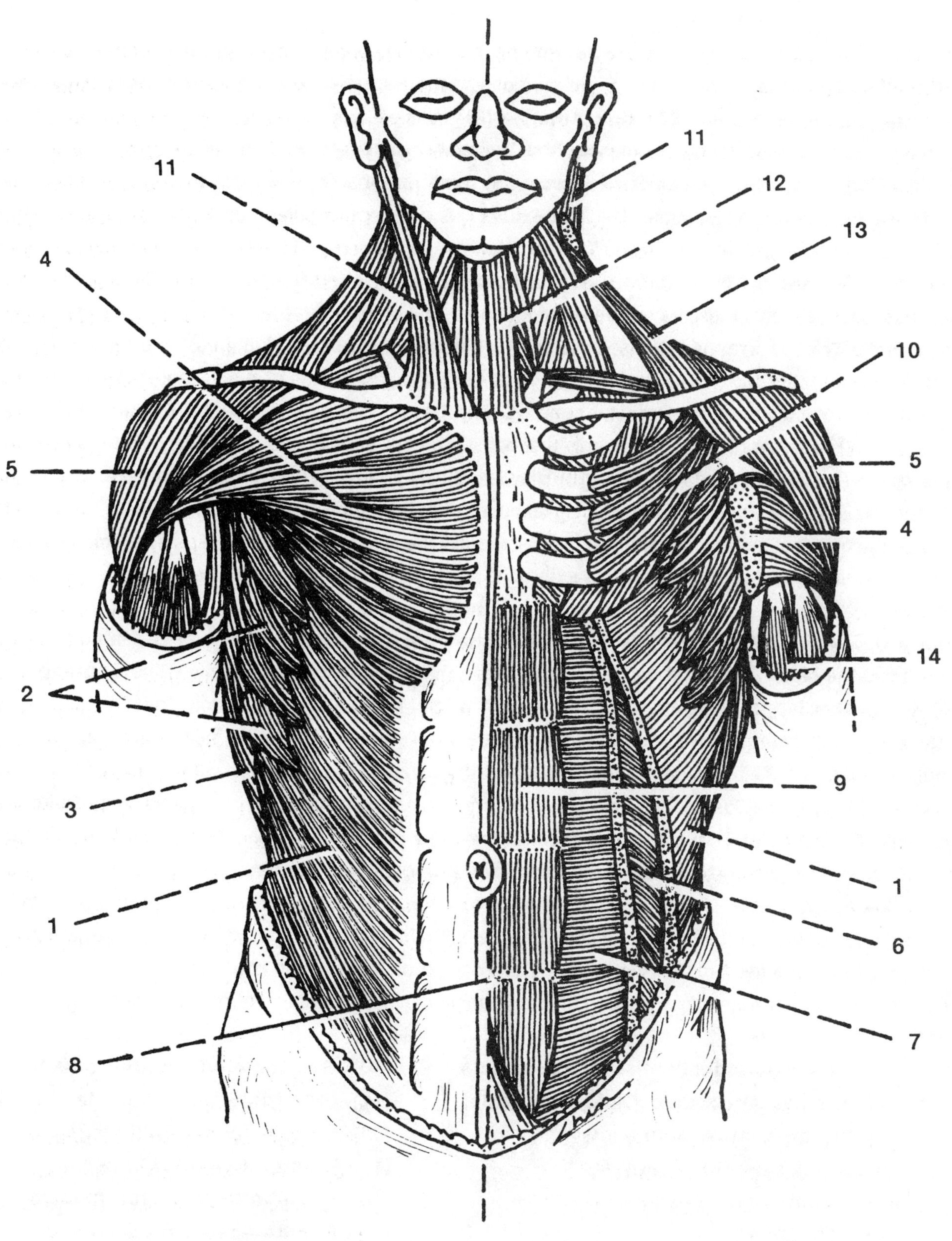
11
11
12
13
4
10
5
5
4
14
2
9
3
1
1
6
8
7

Posterior Muscles of the Trunk

(Superficial and Deep Muscles)

The Superficial Muscles of the Trunk are pictured on the Left side of the body, while the Deeper Muscles are seen on the Right side of the body. Small portions of the **External Abdominal Oblique (1)**, **Internal Abdominal Oblique (2)**, and **Serratus Anterior (3)** muscles are shown lying under the protection of the massive **Latissimus Dorsi (4)** muscle (the "swimmer's muscle") which originates from the lower six Thoracic Vertebrae, the **Lumbodorsal Fascia (5)**, and from the **Crest of the Ilium (6)** to insert upon the Bicipital Groove of the Humerus. The **Trapezius (7)** is a large, triangular muscle that originates from the Occipital Bone and all of the Cervical and Thoracic Vertebrae to insert upon the lateral third of the Clavicle, the Spine of the Scapula and the Acromion. The **Deltoid (8)** muscle originates from the insertion points of the Trapezius and these two muscles are believed by some anatomists to represent a single muscle. A small portion of the **Triceps Brachii (9)** is seen along the dorsal surface of the Humerus. The insertion point of the **Sternocleidomastoid (10)** muscle is shown at the Mastoid Process of the Temporal bone and along the Superior Nuchal line of the Occipital bone. Of the Deeper muscles, notice the **Erector Spinae (11)** muscle which is a series of muscle segments that begin from the Sacrum and Iliac Crest to extend up to base of the skull. The **Semispinalis Capitis (12)** likewise consists of numerous muscular slips extending from the upper six Thoracic Vertebrae to the Occipital bone. The Errector Spinae and Semispinalis are important postural muscles that serve to extend and laterally rotate the spine. Deep to the Trapezius muscle can be seen the **Levator Scapulae (13)** muscle extending from the first four Cervical Vertebrae to the Superior Angle of the Scapula, which acts along with the **Rhomboideus Major (14)** and **Minor (15)** to pull the Scapula Upwards and Medially. The Rhomboids arise from the Vertebral Spines of C-7 through T-5 and insert along the Medial border of the Scapula between the Spine and Inferior angle. The **Supraspinatus (16)** and **Infraspinatus (17)** muscles are shown originating above and below the Scapular Spine, respectively, with both muscles inserting into the Greater Tuberosity of the Humerus and Capsule of the Shoulder Joint. These two muscles act to support the Head of the Humerus within the Glenoid Fossa, providing stability to the Joint during the actions of other shoulder muscles, most especially the **Deltoid (8)**. Two clinically significant Triangles are present in this view of the trunk. The **Triangle of Auscultation (18)** lies above the horizontal border of the Latissimus Dorsi, with the vertebral border of the Scapula along with the lower edge of the Trapezius completing the triangle. The floor of the **Lumbar Triangle** is evident in the exposed portion of the Internal Oblique Muscle, with the triangle bordered by the anterior edge of the Latissimus Dorsi and the Crest of the Ileum. The Triangle of Auscultation is utilized in examination of the chest, esophagus and stomach, while the Lumbar Triangle can be the location of rare lumbar hernias.

(1) External Abdominal Oblique muscle
(2) Internal Abdominal Oblique muscle
(3) Serratus Anterior muscle
(4) Latissimus Dorsi muscle
(5) Lumbodorsal Fascia
(6) Iliac Crest
(7) Trapezius muscle
(8) Deltoid muscle
(9) Triceps Brachii muscle
(10) Sternocleidomastoid muscle
(11) Errector Spinae muscle
(12) Semispinalis Capitis muscle
(13) Levator Scapulae muscle
(14) Rhomboideus Major muscle
(15) Rhomboideus Minor muscle
(16) Supraspinatus muscle
(17) Infraspinatus muscle
(18) Triangle of Ascultation

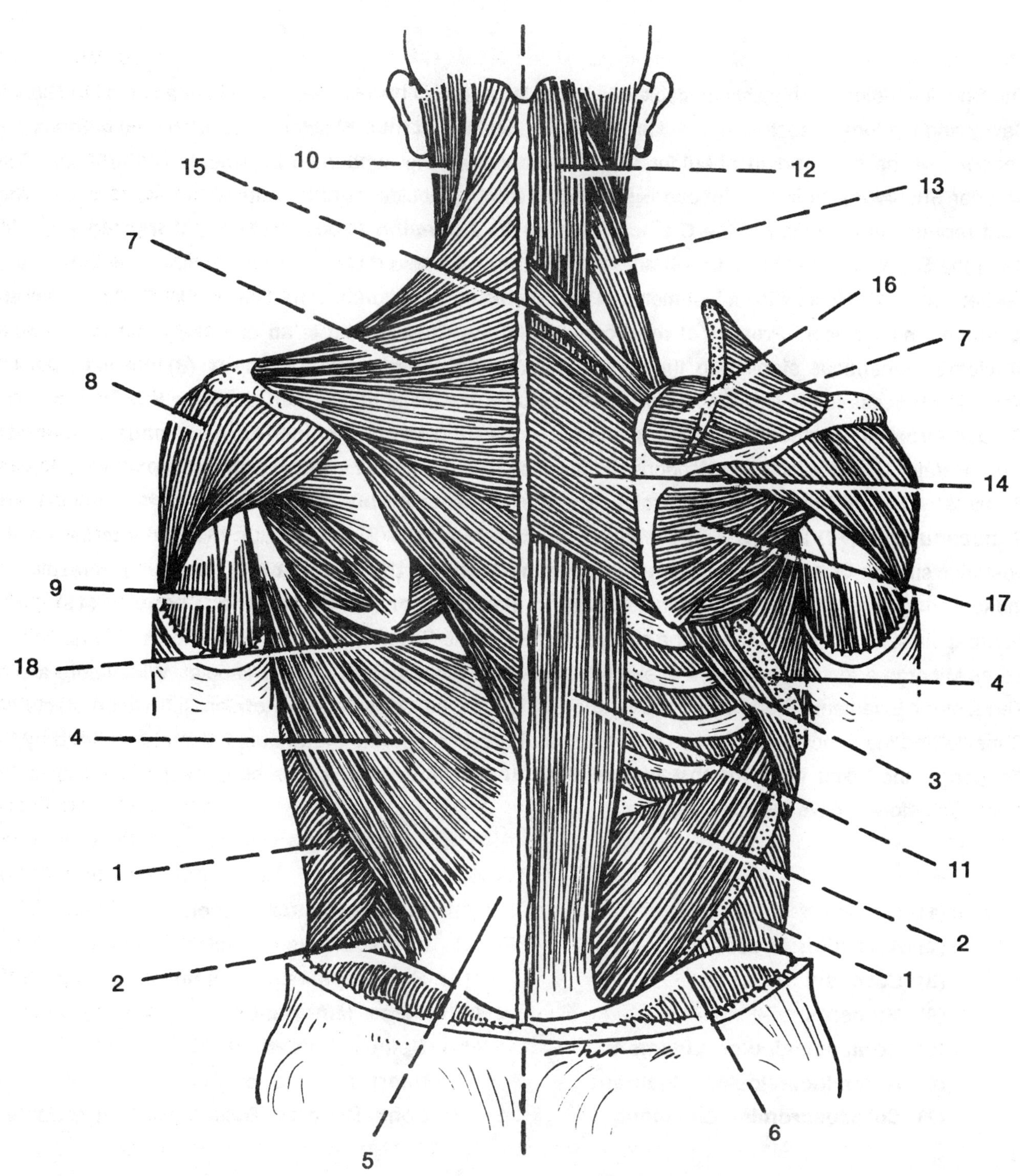

MS-6

Shoulder Joint: Ligaments and "Rotator Cuff" Muscles

(Anterior and Posterior Views)

The Shoulder Joint is somewhat unstable due to the large size of the Humeral Head in relation to the Glenoid Cavity and the loose structure of the Joint Capsule. Stability of the Shoulder is provided by the Tonus and Tendons of the surrounding Shoulder Muscles as well as the supportive Ligaments of the Joint. The Anterior Shoulder joint is seen above, while the Posterior Shoulder joint is pictured below. The Shoulder Joint represents the union of the **Clavicle (1)** with the **Acromion Process (2)** and **Corocoid Process (3)** of the Scapula. The **Humerus (4)** articulates with the Glenoid Cavity to form a Synovial lined Ball-and-Socket Joint. Three significant Ligaments of the Shoulder are pictured. The **Coracoclavicular Ligament (5)** unites the Corocoid Process of the Scapula with the Clavicle and is an extremely strong structure, providing tremendous stability to the joint. The **Acromioclavicular Ligament (6)** passes from the Acromion Process to the Clavicle, helping to firmly unite the distal end of the Clavicle to the Scapula. The **Coracoacromial Ligament (7)** acts to prevent the upward dislocation of the Humeral Head. The surgical term "Rotator Cuff" refers to the Tendons of four shoulder muscles which fuse with the lateral portion of the Capsule. These Rotator Cuff muscles are the **Supraspinatus (8)**, the **Infraspinatus (9)**, the **Subscapularis (10)** and the **Teres Minor (11)** muscles. The Supraspinatus muscle is found on the posterior surface of the Scapula, situated above the **Spine of the Scapula (12)**, while the Infraspinatus muscle originates below the Spine, down to the **Inferior Angle of the Scapula (13)**. The Subscapularis muscle is located on the anterior surface of the Scapula, filling the Subscapular Fossa, and the Teres Minor is a rounded muscle that extends from the Axillary Border of the Scapula. While there are no Tendons or Ligaments found along the inferior margin of the Capsule, the Rotator Cuff well protects the Capsule and the Synovial Membrane from trauma. Further support to the Shoulder Joint is provided by the Tendon of the **Long Head of the Biceps Brachii (14)** muscle which crosses over the Head of the Humerus before it leave the Joint Capsule.

(1) Clavicle
(2) Acromion Process
(3) Corocoid Process
(4) Humerus
(5) Coracoclavicular Ligament
(6) Acromioclavicular Ligament
(7) Coracoacromial Ligament
(8) Supraspinatus muscle
(9) Infraspinatus muscle
(10) Subscapularis muscle
(11) Teres Minor muscle
(12) Spine of Scapula
(13) Inferior Angle of Scapula
(14) Long Head of Biceps Brachii muscle

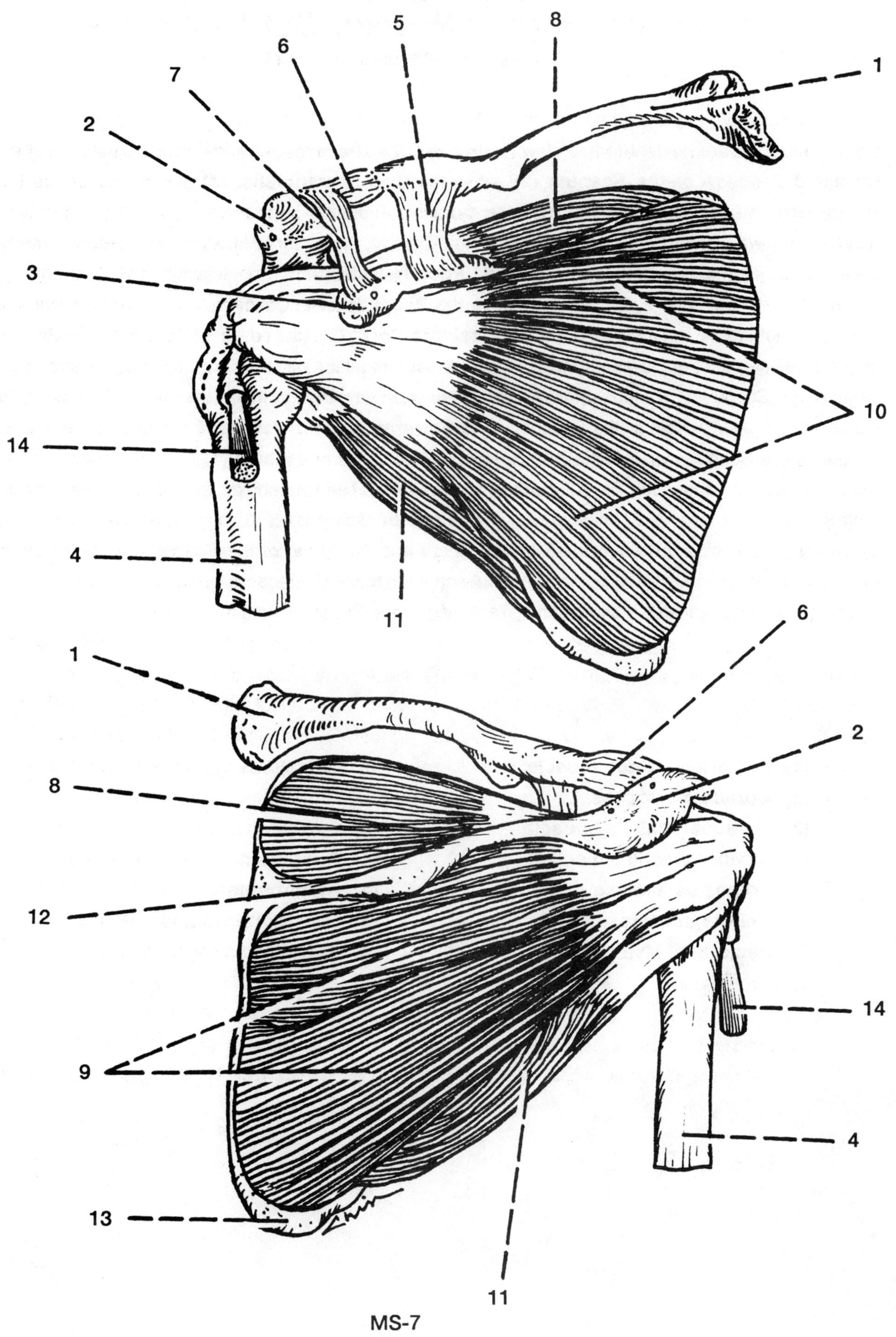

MS-7

Anterior Shoulder Muscles

(Right side)

The Right Shoulder is pictured here from the Ventral or Anterior direction. Common Origins and Inserts of several muscles are evident in this view. Notice that the **Pectoralis Minor (1)** muscle Inserts upon the **Coracoid Process of the Scapula (2)**, while the **Coracobrachialis (3)** and the **Short Head** of the **Biceps Brachii (4)** both Originate from the Coracoid Process. A small segment of the **Trapezius (5)** muscle is shown Inserting on the lateral third of the Clavicle, in the same location that serves as the Origin of the **Deltoid (6)** muscle. Notice also that the **Pectoralis Major (7)** muscle must cross the belly of the Biceps Brachii muscle to eventually Insert into the **Bicipital** (or Intertubercular) **Groove of the Humerus** where it is joined by the Insertion of the **Latissimus Dorsi (8)** and **Teres Major (9)** muscles. Note that only a small segment of the Teres Major muscle is evident as it passes forward from its origin at the inferior angle of the Scapula . Various relationships of common Origins and Insertions are present not only at the Humerus, but also pertain to the Scapula. The **Serratus Anterior (10)** inserts along the entire Vertebral Border of the Scapula, while the **Rhomboideus Major (11)** and **Minor (12)** also insert at the Vertebral Border, below and above the Scapular Spine respectively. (The Rhomboids were given their name because their shape is similar to a parallelogram.) The **Levator Scapulae (13)** also inserts along the Vertebral border of the Scapula, between the Superior Angle and the Spine of the Scapula. The **Subscapularis (14)** muscle is seen Originating from the Subscapular fossa to become one of the four "Rotator Cuff" muscles (the Supraspinatus, Infraspinatus, Teres Minor and Subscapularis make up this group).

(1) Pectoralis Minor muscle
(2) Coracoid Process
(3) Coracobrachialis muscle
(4) Biceps Brachii muscle
(5) Trapezius muscle
(6) Deltoid muscle
(7) Pectoralis Major
(8) Latissimus Dorsi muscle
(9) Teres Major muscle
(10) Serratus Anterior muscle
(11) Rhomboideus Major muscle
(12) Rhomboideus Minor muscle
(13) Levator Scapulae muscle
(14) Subscapularis muscle

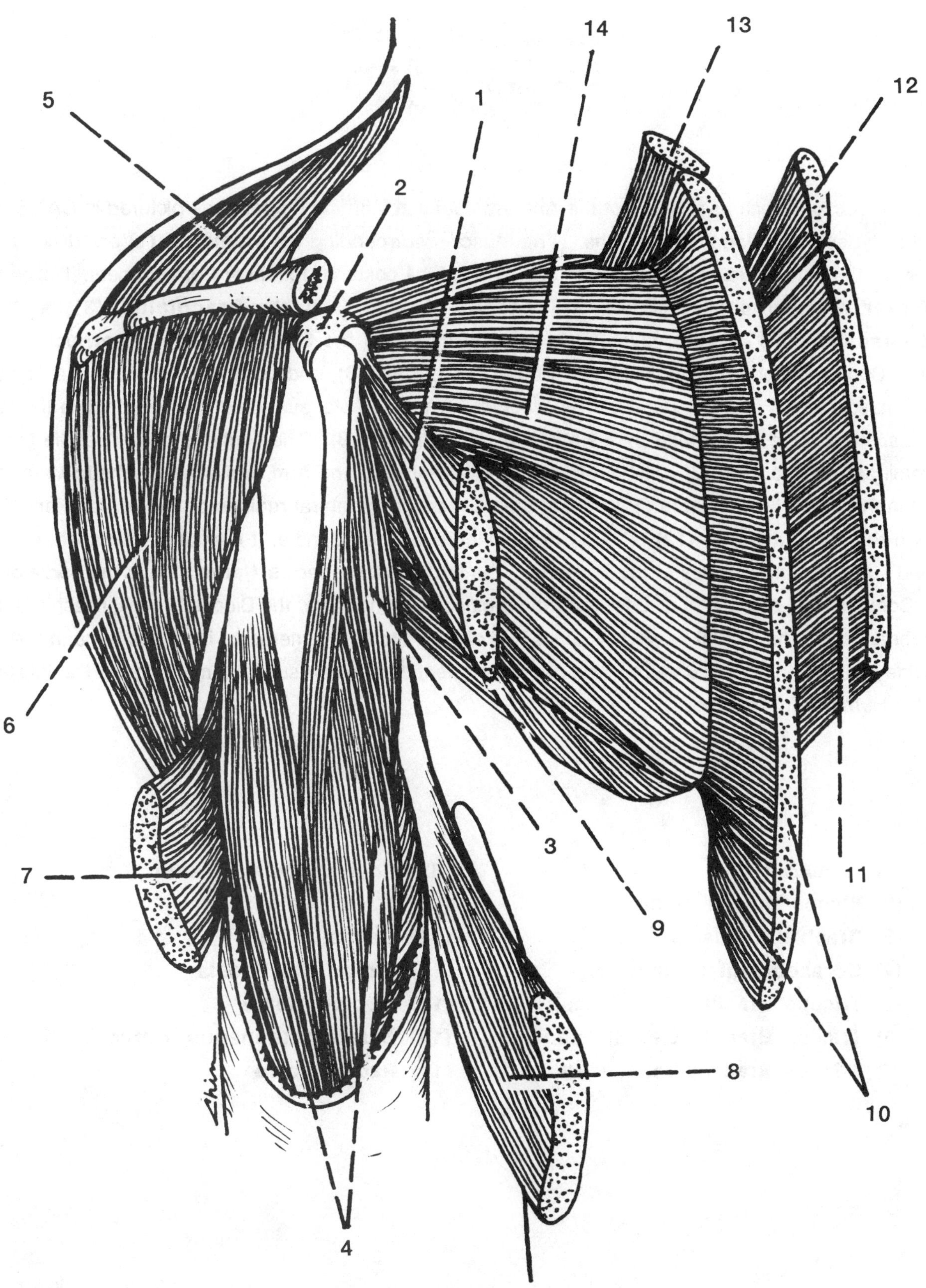

MS-8

Cross Section of the Right Arm

(Inferior View)

Transverse sections such as this one of the Arm are useful in learning the features pictured in CAT Scans and other types of sectional radiographs. The Muscles surrounding the **Humerus (1)** are divided into those within the ventral Flexor Compartment and those of the posterior Extensor Compartment. Notice that the Flexor Compartment includes the **Biceps Brachii (2)**, the **Brachialis (3)**, and the **Coracobrachialis (4)** muscles which all function in flexion of the Shoulder and/or Elbow Joint. The Extensor Compartment is occupied by the **Medial (5)**, **Lateral (6)**, and **Long (7) Heads** of the **Triceps Brachii** muscle which acts to extend the Elbow Joint. On the Medial surface of the Arm, notice the large Neurovascular bundle surrounded by a **Neurovascular Sheath (8)**. The contents of this bundle include the prominent **Basilic Vein (9)** which drains the medial Forearm and Arm, while the **Cephalic Vein (10)**, located in the **Superficial Fascia (11)**, drains the corresponding lateral regions of the Forearm and Arm. Although the **Median Nerve (12)** is found in this Neurovascular Bundle, it contains no branches in the Arm and supplies only muscles of the Forearm. The **Musculocutaneous Nerve (13)** is the nerve of the Flexor Compartment of the Arm, supplying all three muscles found there: the Biceps Brachii, Brachialis, and Coracobrachialis. The **Radial nerve (14)** is located adjacent to the Humerus as it passes through the Arm to provide innervation in the Forearm. This particular orientation may result in damage to the Radial Nerve when the Humerus in fractured.

(1) Humerus
(2) Biceps Brachii muscle
(3) Brachialis muscle
(4) Coracobrachialis muscle
(5) Triceps Brachii, Medial head
(6) Triceps Brachii, Lateral head
(7) Triceps Brachii, Long head
(8) Neurovascular Sheath
(9) Basilic Vein
(10) Cephalic Vein
(11) Superficial Fascia
(12) Median Nerve
(13) Musculocutaneous Nerve
(14) Radial Nerve

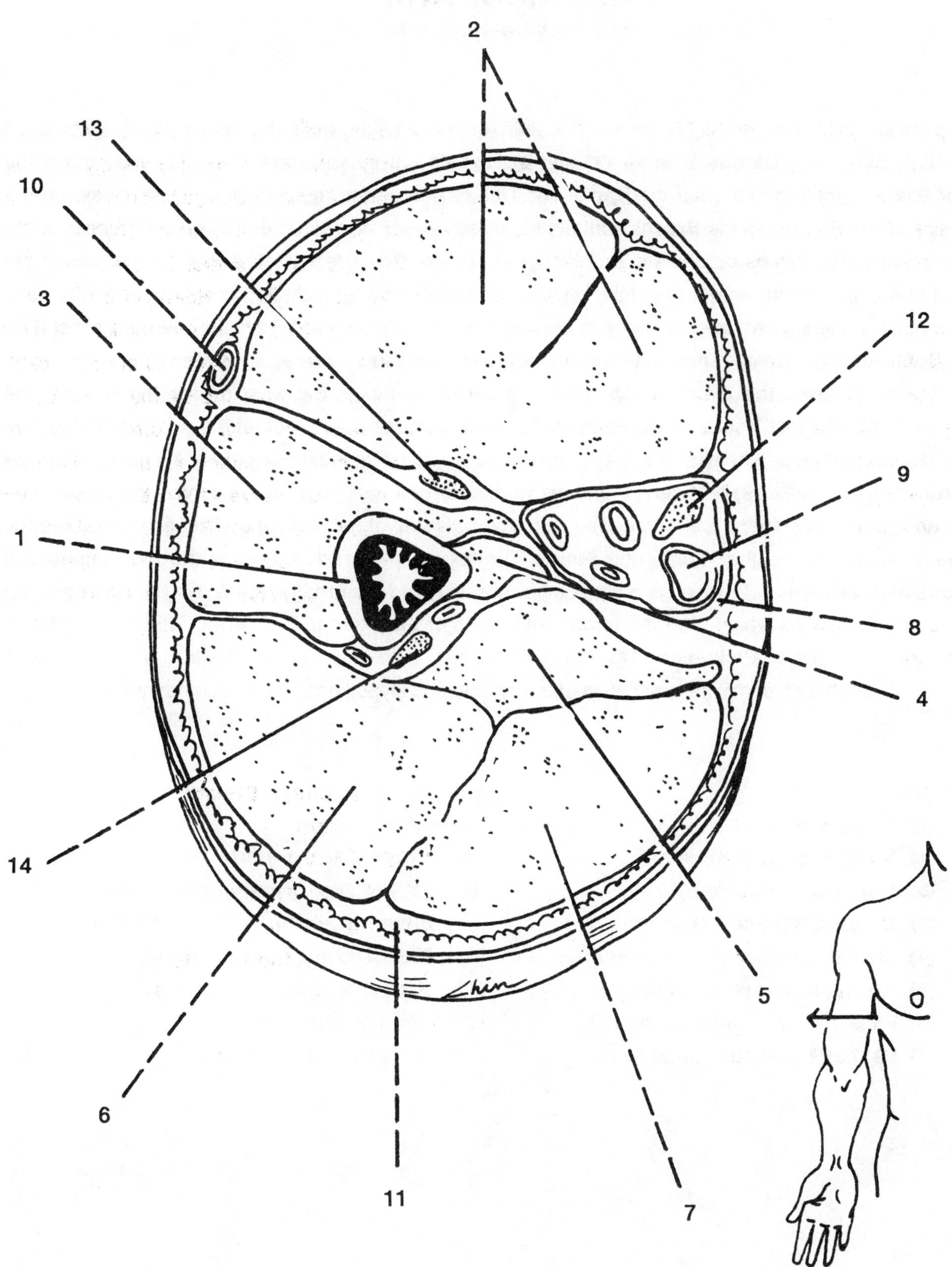
2
13
10
3
12
9
1
8
4
14
5
6
11
7
Chin

Superficial and Deep Forearm Muscle

(Anterior View, Right side)

The Superficial Muscles of the Right Forearm are seen on the Left side, while the Deep Muscle are pictured on the Right side. The **Biceps Brachii (1)** and **Brachialis (2)** muscles are shown as they cross the Anterior Elbow Joint to insert upon the Radial and Ulnar Tuberosities, respectively. Descending on the lateral side of the Forearm is the **Brachioradialis (3)** muscle which inserts upon the Styloid Process of the Radius, assisting the Biceps Brachii and Brachialis in flexion of the Forearm. The intrinsic muscles of the Forearm and extrinsic muscles of the Hand display common features, including common sites of Origin. There are Five Superficial muscles involved in Flexion/Pronation that all arise from the **Medial Epicondyle of the Humerus (4)**. Named from lateral to medial, these are the rounded **Pronator Teres (5)** which rotates the palm down, the **Flexor Carpi Radialis (6)** which flexes the wrist on the thumb side, the **Palmaris Longus (7)** that inserts on the Palmar Aponeurosis to assist in flexion of the wrist and tenses the palm, the **Flexor Carpi Ulnaris (8)** which flexes the wrist on the little finger side and the **Flexor Digitorum Superficialis (9)** that sends tendons to each of the four fingers and serves to flex both the fingers and the wrist. There are Three Deep Muscles also involved in Flexion/Pronation: the **Flexor Pollicis Longus (10)** to the Thumb, the **Flexor Digitorum Profundus (11)** to the four fingers and the **Pronator Quadratus (12)** which is a flat square muscle found near the wrist. Within the Hand are two sets of Intrinsic muscles which form the prominent bulges of the **Thenar Eminence (13)** on the Thumb side and the **Hypothenar Eminence (14)** on the Little finger side of the palm. Refer to the examination of the Palm (MS-12) for a complete description of the Thenar and Hypothenar muscle components.

(1) Biceps Brachii muscle
(2) Brachialis muscle
(3) Brachioradialis muscle
(4) Medial Epicondyle of Humerus
(5) Pronator Teres muscle
(6) Flexor Carpi Radialis muscle
(7) Palmaris Longus muscle
(8) Flexor Carpi Ulnaris muscle
(9) Flexor Digitorum Superficialis
(10) Flexor Pollicis Longus muscle
(11) Flexor Digitorum Profundus
(12) Pronator Quadratus muscle
(13) Thenar muscles
(14) Hypothenar muscles

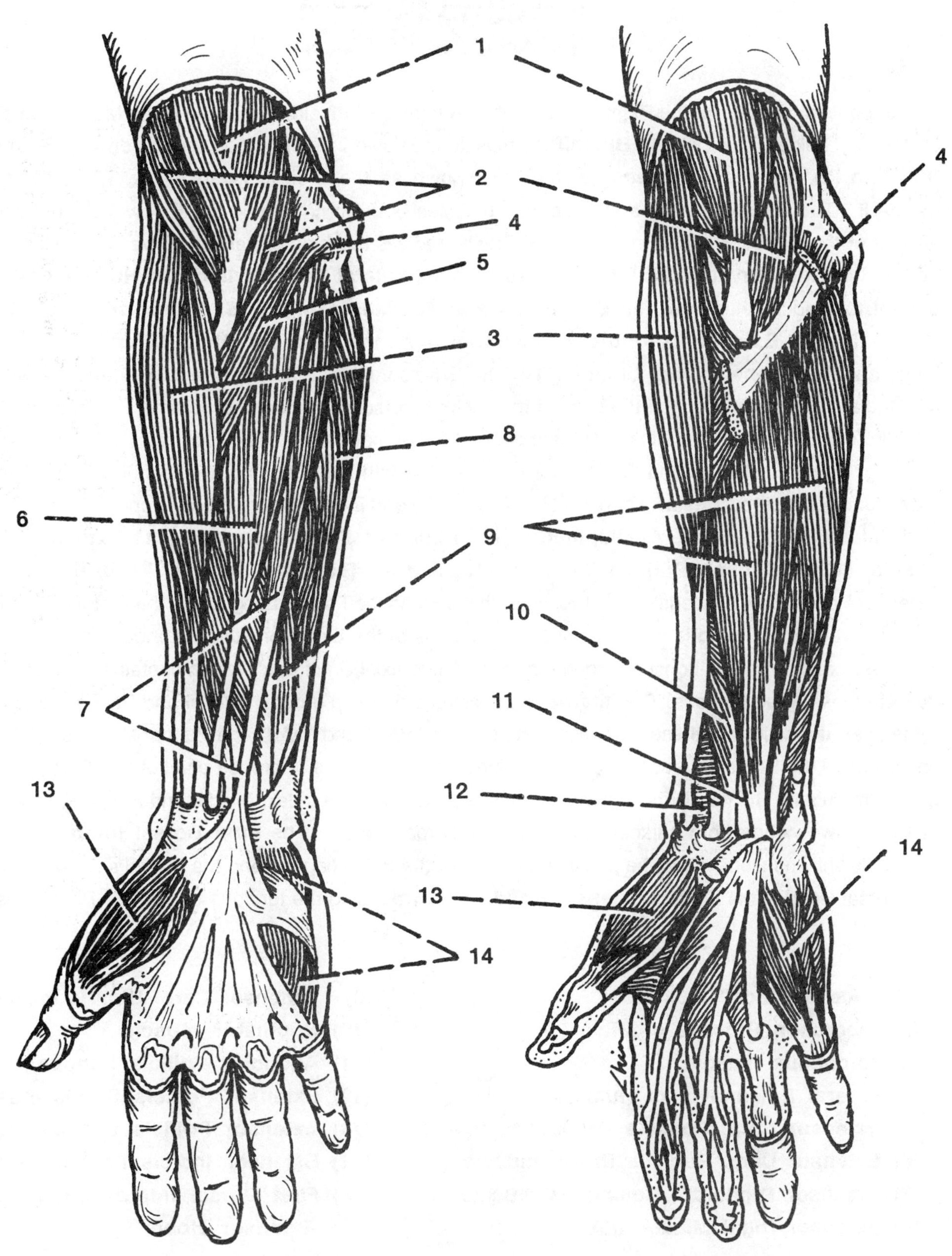

MS-10

Superficial and Deep Forearm Muscle

(Posterior View, Right side)

The Superficial Muscles of the Right Forearm are seen on the Left side, while the Deep Muscle are pictured on the Right side. The **Triceps Brachii (1)** muscle is shown inserting upon the **Olecranon Process of the Ulna (2)**, with the **Anconeus (3)** muscle (known as the "Fourth Head of the Triceps") inserting beside it. The Triceps Brachii and Anconeus muscles both act to extend the Elbow joint, opposing the actions of the Biceps Brachii, Brachialis and Brachioradialis muscles. There are Five Superficial muscles involved in Extension/Supination that all arise from the **Lateral Epicondyle of the Humerus (4).** From Lateral to Medial, these are the **Extensor Carpi Radialis Longus (5), Extensor Carpi Radialis Brevis (6)**, the **Extensor Digitorum Communis (7)**, the **Extensor Digiti Minimi (8)** to the little finger and the **Extensor Carpi Ulnaris (9)**. The Extensor Carpi muscles all act to extend the wrist, with the Radialis Longus inserting at the base of the second metacarpal bone, the Radialis Brevis at the base of the third metacarpal and the Ulnaris at the base of the fifth metacarpal. The Extensor Digitorum Communis and Digiti Minimi act to extend the four fingers, with the fifth finger having two tendons. The Five Deep Extensor/Supinator muscles act upon the Thumb (Pollicis) and Index Finger (Indicis Proprius). From proximal to distal in origin, these are the **Supinator (10), Abductor Pollicis Longus (11), Extensor Pollicis Brevis (12), Extensor Pollicis Longus (13)** and the **Extensor Indicis Proprius (14)**. The Extensor Pollicis Brevis inserts on the Proximal Phalanx and the Pollicis Longus into the Distal Phalanx of the thumb. When the Thumb is fully extended, the tendons of the Extensor Pollicis Longus and Brevis, along with the Abductor Pollicis Longus tendon form the "Anatomical Snuff Box", which contains the Radial Artery deep to these three tendons. The **Interossei** muscles are in two groups, Dorsal and Palmar. It is easy to remember the actions of these muscles by the formula "PAD" and "DAB": the Palmar Interossei Adduct the fingers while the Dorsal Interossei Abduct the fingers relative to the reference point of the Third metacarpal and Third finger. The four Dorsal Interossei muscles extend to the Index, Middle and Ring fingers and each arises as two heads from adjacent sides of Metacarpal bones. The **First Dorsal Interosseous (15)** muscle is shown inserting into the proximal phalanx of the Index finger along the Radial side. Also seen on the Dorsal Hand is the Fibrous **Extensor Hood (16)** surrounding the tendons to the four Phalanges.

(1) Triceps Brachii muscle
(2) Olecranon Process of Ulna
(3) Anconeus muscle
(4) Lateral Epicondyle of Humerus
(5) Extensor Carpi Radialis Longus muscle
(6) Extensor Carpi Radialis Brevis muscle
(7) Extensor Digitorum Communis muscle
(8) Extensor Digiti Minimi muscle
(9) Extensor Carpi Ulnaris muscle
(10) Supinator muscle
(11) Abductor Pollicis Longus muscle
(12) Extensor Pollicis Brevis muscle
(13) Extensor Pollicis Longus muscle
(14) Extensor Indicis Proprius muscle
(15) First Dorsal Interosseous muscle
(16) Extensor Hood

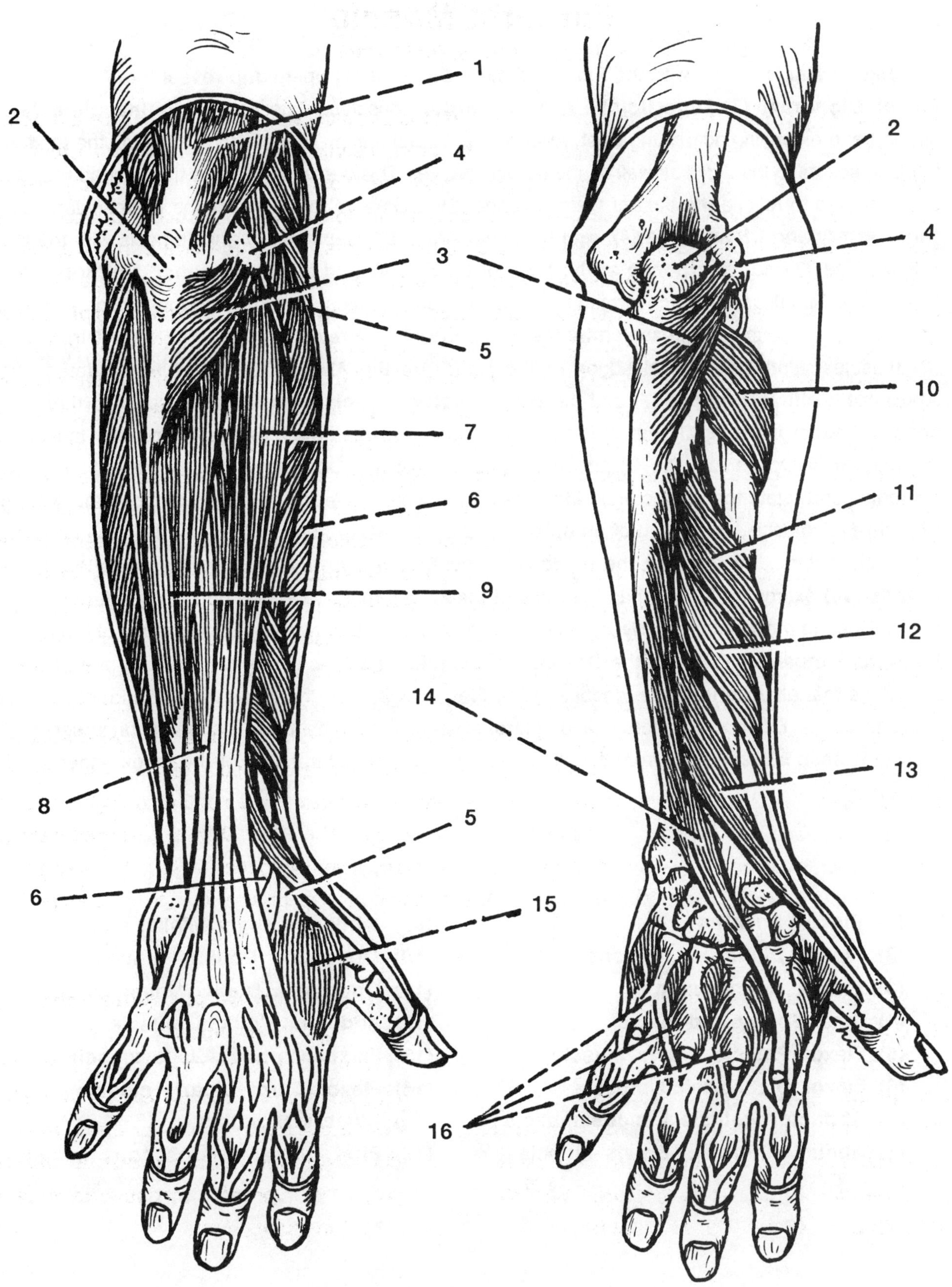

MS-11

Muscles of the Palm

(Right side)

The Right Palm is shown with the Fibrous **Flexor Sheaths (1)** opened to reveal the Tendons of the **Flexor Digitorum (2) Superficialis** and **Profundus** muscles. The **Flexor Retinaculum (3)** is a strong band extending across the wrist, forming the Carpal Flexor Tunnel which contains the tendons of the long flexors to the digits as well as the Median Nerve. This dense connective tissue wrap may be cut and resealed to relieve the pain of Carpal Tunnel Syndrome. Crossing over the Retinaculum are the **Ulnar Artery and Ulnar Nerve (4)**, with the Ulnar Artery later forming an anastomosis with the Radial Artery. The **Thenar Eminence** at the base of the Thumb and the **Hypothenar Eminence** at the base of the Fifth finger are each formed by three categories of muscle: closest to the midline of the Hand are the Flexor muscles represented by the **Flexor Pollicis Brevis (5)** and the **Flexor Digiti Minimi (6)** muscles; along the lateral edges of the Hand are the Abductor muscles represented by the **Abductor Pollicis Longus (7)** and the cut **Abductor Pollicis Brevis (8)** muscles to the Thumb, and the **Abductor Digiti Minimi (9)** to the Fifth finger. Between the Flexor and Abductor muscles are the muscle of Opposition, the **Opponens Pollicis (10)** and **Opponens Digiti Minimi (11)**, which act to flex and rotate the Thumb and Fifth finger. Assisting in opposition of the Thumb is the **Adductor Pollicis (12)** muscle, composed of an oblique and a transverse head, which insert on the base of the proximal phalanx. Notice that the tendons of the **Flexor Carpi Radialis (13)** and **Flexor Carpi Ulnaris (14)** extend no further than the base of the Metacarpal bones and do not pass further into the Palm. Deep to the Thenar and Hypothenar muscles can be found two additional groups of muscles, the **Lumbrical muscles (15)** and the **Dorsal Interossei**. The four Lumbrical muscles are "wormlike" structures that all arise from the tendons of the Flexor Digitorum Profundus muscle and insert into the tendons of the Extensor Digitorum and Interossei muscles. The **First Dorsal Interosseous (16)** muscle is seen bulging away from the thumb to insert on the lateral side of the Index finger proximal phalanx.

(1) Flexor Sheaths
(2) Flexor Digitorum tendons
(3) Flexor Retinaculum
(4) Ulnar Artery and Nerve
(5) Flexor Pollicis Brevis muscle
(6) Flexor Digiti Minimi muscle
(7) Abductor Pollicis Longus muscle
(8) Abductor Pollicis Brevis muscle
(9) Abductor Digiti Minimi muscle
(10) Opponens Pollicis muscle
(11) Opponens Digiti Minimi muscle
(12) Adductor Pollicis muscle
(13) Flexor Carpi Radialis muscle
(14) Flexor Carpi Ulnaris muscle
(15) Lumbrical muscles
(16) First Dorsal Interosseous muscle

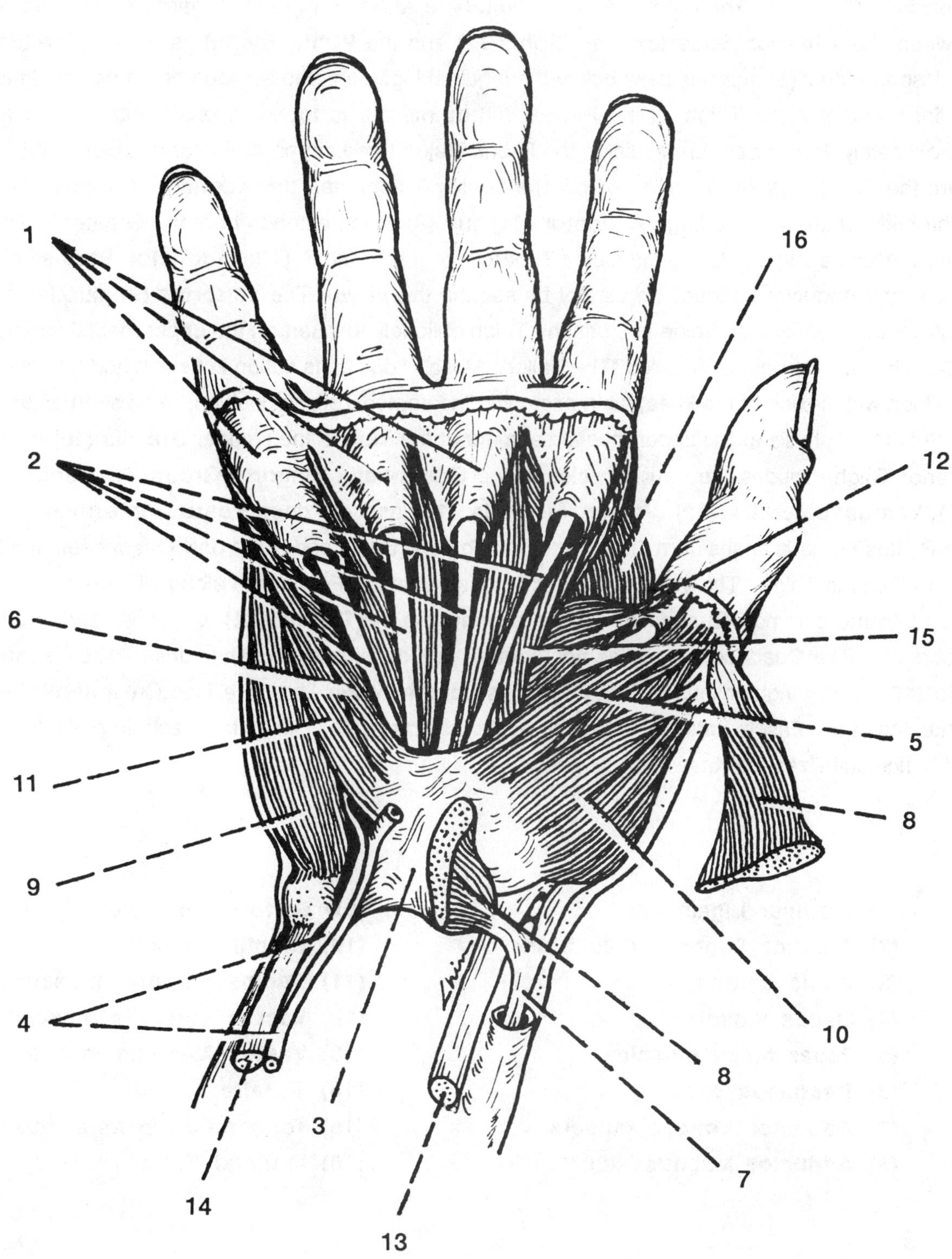
1
16
2
12
6
15
5
11
8
9
4
10
8
3
7
14
13

Superficial Anterior Thigh Muscles

(Right side)

The Superficial Anterior muscles of the Thigh are principally concerned with Flexion of the Hip and Extension of the Leg. The Right side is pictured here with the **Inguinal Ligament (1)** seen extending between the **Anterior Superior Iliac Spine (2)** and the **Pubic Symphysis (3)**. The **Iliacus (4)** and **Psoas Major (5)** muscles pass below the Inguinal Ligament and act together in powerful flexion and medial rotation of the Thigh. (The **Psoas Minor** muscle is found inconsistently, being absent in approximately forty percent.) Medial to the Psoas Major, three strong Adductor muscles of the Thigh are seen: the **Pectineus (6)** inserting below the Lesser Trochanter, the **Adductor Longus (7)** inserting at the mid-Femur, and the large **Adductor Magnus (8)** which extends from the Greater Trochanter all along the Linea Aspera to the Adductor Tubercle of the Femur. (The **Adductor Brevis** muscle lies deep to the Adductor Longus and cannot be seen in this view.) The **Sartorius (9)** muscle arises from the Anterior Superior Iliac Spine to cross the Thigh obliquely to insert on the upper medial segment of the Tibia. This muscle has been called "The Tailor's Muscle" due to its action in flexion and lateral rotation of the Thigh with flexion of the knee, as in assuming the seated Tailor's position. Also seen extending from the Pubic Symphysis to the upper medial segment of the Tibia is the slender **Gracilis (10)** muscle. The Anterior Thigh includes the Four muscles of the **Quadriceps Femoris Group**: the **Rectus Femoris (11)**, **Vastus Lateralis (12)**, **Vastus Medialis (13)** and the deep **Vastus Intermedius** muscle. The Rectus Femoris arises from the Anterior Inferior Iliac Spine and is the only Quadriceps muscle which acts to Flex the Thigh. The Vastus muscles all arise from the Femur and along with the Rectus Femoris insert into the common Quadriceps tendon suspending the **Patella (14)** to finally insert into the Tibial Tuberosity. The Quadriceps muscles act together in Extension of the Knee Joint. The **Tensor Fascia Lata (15)** is a triangular muscle that originates from the outer lip of the Iliac Crest and inserts into a thickening of the Fascia Lata known as the **Iliotibial Tract (16)**. This muscle acts to pull upon the Tract and hence stabilizes the Knee Joint.

(1) Inguinal Ligament
(2) Anterior Superior Iliac Spine
(3) Pubic Symphysis
(4) Iliacus muscle
(5) Psoas Major muscle
(6) Pectineus muscle
(7) Adductor Longus muscle
(8) Adductor Magnus muscle
(9) Sartorius muscle
(10) Gracilis muscle
(11) Rectus Femoris muscle
(12) Vastus Lateralis muscle
(13) Vastus Medialis muscle
(14) Patella
(15) Tensor Fascia Lata muscle
(16) Iliotibial Tract

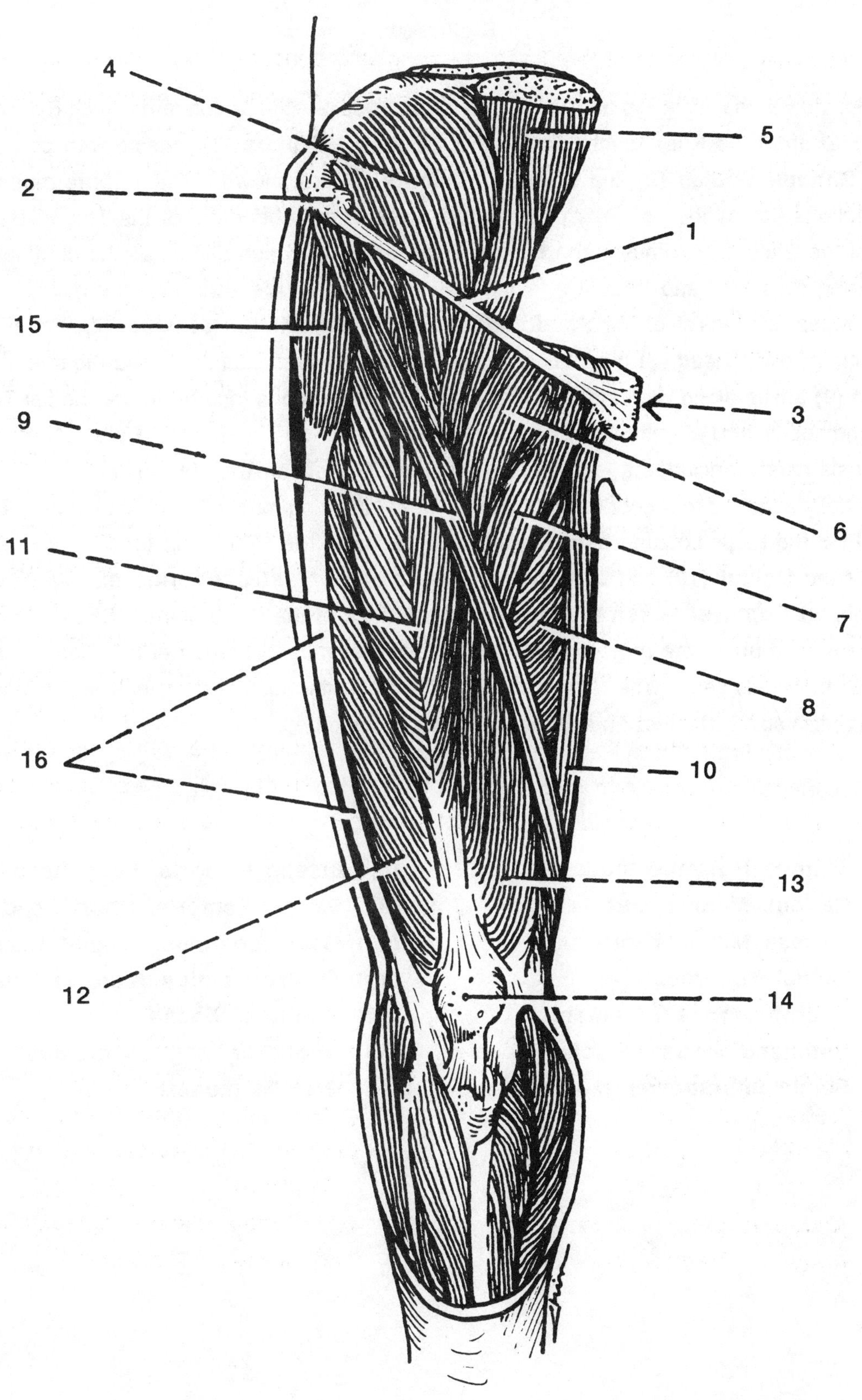
4
5
2
1
15
3
9
6
11
7
8
16
10
13
12
14

Superficial Posterior Thigh Muscles

(Right side)

This Posterior view of the Right Hip and Thigh reveals muscles principally concerned with Extension of the Thigh and Flexion of the Knee Joint. The entire **Gluteus Maximus (1)** muscle with portions of the underlying **Gluteus Medius (2)** and **Gluteus Minimus (3)** are shown. These three muscles run in various directions across the posterior pelvis to act as powerful Abductors of the Thigh. Beneath the protection of the Gluteus Maximus is the **Ischial Tuberosity (4)** which serves as the common origin of the "Hamstring Muscles" and attaches to the powerful **Sacrotuberous Ligament (5)**. The three "Hamstring Muscles" consist of the **Semitendinosus (6)**, **Semimembranosus (7)**, and the **Biceps Femoris** with its **Long Head (8)** originating off the Tuberosity and Sacrotuberous Ligament, while the **Short Head (9)** arises along the Linea Aspera. The Biceps Femoris inserts on the medial Tibia, while the Semitendinosus and Semimembranosus muscles insert on the lateral Tibia. Notice that the Semitendinosis muscle includes a long stout tendon, while the Semimembranosus muscle contains a large fleshy belly attached to a shortened tendon. The "Hamstring muscles" act in unison to Extend the Thigh and Flex the Leg. Located between the divergence of the Hamstring tendons above, and the beginning of the **Medial (10)** and **Lateral (11) Heads** of the **Gastrocnemius** muscle below, is the diamond-shaped **Popliteal Fossa (12)** located at the back of the Knee Joint. Situated deep to the Biceps Femoris and under the protection of the Gluteus Maximus, a portion of the massive **Adductor Magnus (13)** muscle is seen. The **Gracilis (14)** muscle assists the Adductor muscles in movement of the Thigh and also acts in the Flexion and Medial Rotation of the Leg.

(1) Gluteus Maximus muscle
(2) Gluteus Medius muscle
(3) Gluteus Minimus muscle
(4) Ischial Tuberosity
(5) Sacrotuberous Ligament
(6) Semitendinosus muscle
(7) Semimembranosus muscle
(8) Biceps Femoris, Long head
(9) Biceps Femoris, Short head
(10) Gastrocnemius, Medial head
(11) Gastrocnemius, Lateral head
(12) Popliteal Fossa
(13) Adductor Magnus muscle
(14) Gracilis muscle

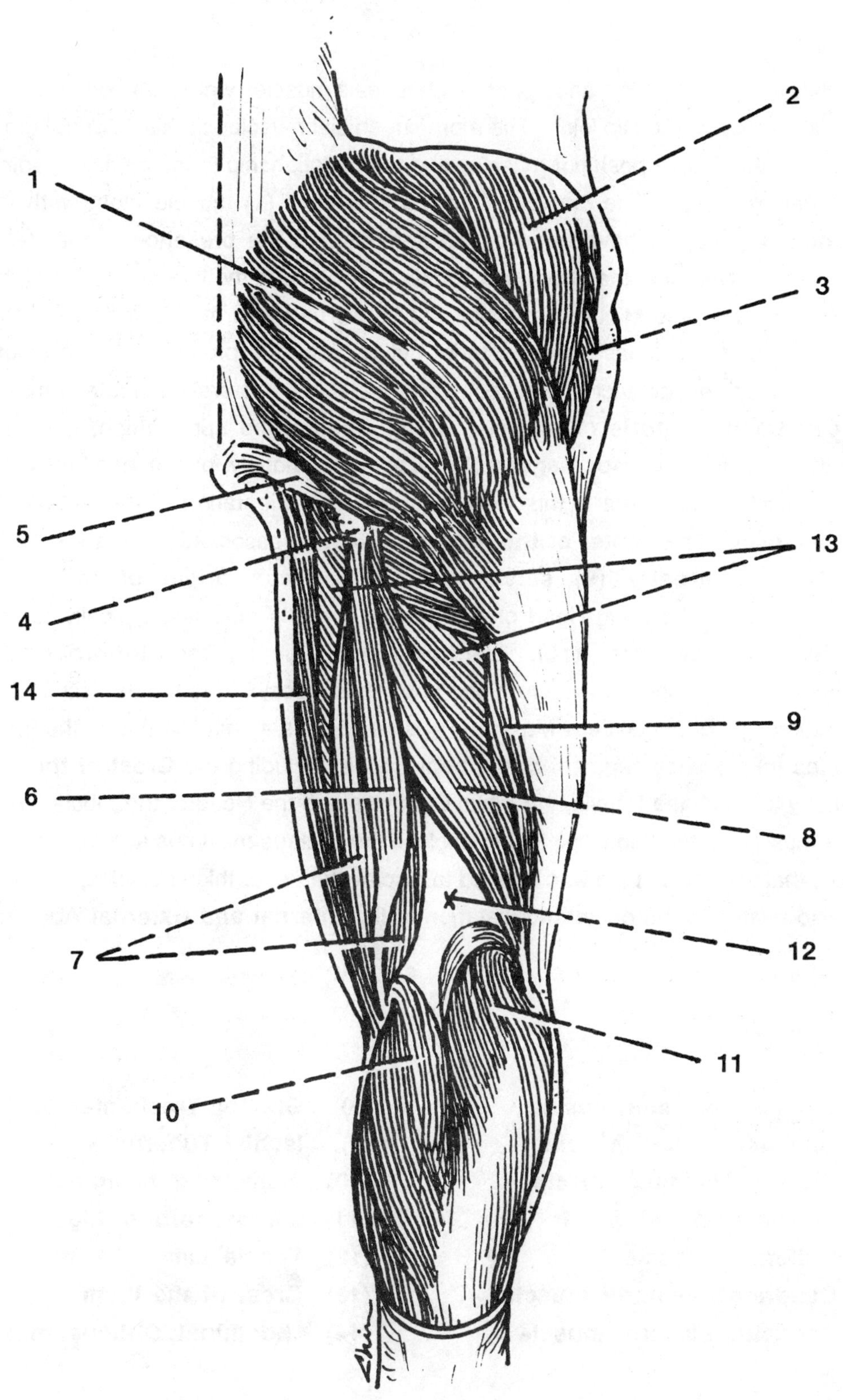

MS-14

Deep Posterior Hip Muscles

(Right side)

This Posterior view of the Right Hip and upper Thigh reveals muscles which are concerned with either the lateral or medial rotation of the Hip joint. The more superficial muscles in this powerful group have been dissected to reveal the deeper posterior muscles and their relationship to the emerging spinal innervation serving the lower extremity. The massive **Gluteus Maximus (1)** muscle along with the underlying **Gluteus Medius (2)** muscle have been cut to demonstrate the presence of the deeper **Gluteus Minimus (3)** muscle. Note that the Gluteus Maximus is involved in the lateral rotation of the thigh, while both the Gluteus Medius and Minimus muscles are concerned with medial rotation of the thigh. The important **Sciatic Nerve (4)** crosses the posterior pelvis under the protection of the stout **Piriforis (5)** muscle and is shielded from contact with the bone by several of the lateral thigh rotator muscles, including the square **Quadratus Femoris (6)** muscle. Upon entering the upper thigh, this nerve is further protected by the massive **Adductor Magnus (7)** muscle. (Obviously, only a small proximal segment of this important muscle is apparent in this view.) Notice the orientation of these powerful hip rotator muscles to the **Greater Trochanter of the Femur (8)** with its associated bursa shown. Realize also that the **Ischial Tuberosity (9)** serves as the common origin of the Semitendinosis, Semimembranosus, and the Long Head of the Biceps Femoris muscles, typically referred to as the combined **"Hamstring Muscles" (10).** The tremendously strong **Sacrotuberous Ligament (11)** extends between the posterior ilium, sacrum and uppermost coccyx to reach the Ischial Tuberosity, and contributes to the origin of the Gluteus Maximus. Also evident is a small portion of the **Fascia Lata (12)** which originates from a wide segment of the bony pelvis, including the **Crest of the Ilium (13)**, and extends inferiorly to reach the Patella, the Tibial Condyles and the Head of the Fibula. The Fascia Lata surrounds the muscles of the thigh, splitting to enclose the Gluteus maximus and the Tensor Fascia Lata muscles. Note that the Fascia Lata was opened to display the upper thigh muscles, and that the Crest of the Ilium is also related to the origin and insertion of the **Internal and External Abdominal Oblique muscles (14)**, respectively.

(1) Gluteus Maximus muscle
(2) Gluteus Medius muscle
(3) Gluteus Minimus muscle
(4) Sciatic Nerve
(5) Piriformis muscle
(6) Quadratus Femoris muscle
(7) Adductor Magnus muscle
(8) Greater Trochanter of Femur
(9) Ischial Tuberosity
(10) "Hamstring muscles"
(11) Sacrotuberous Ligament
(12) Fascia Lata
(13) Crest of the Ilium
(14) Abdominal Oblique muscles

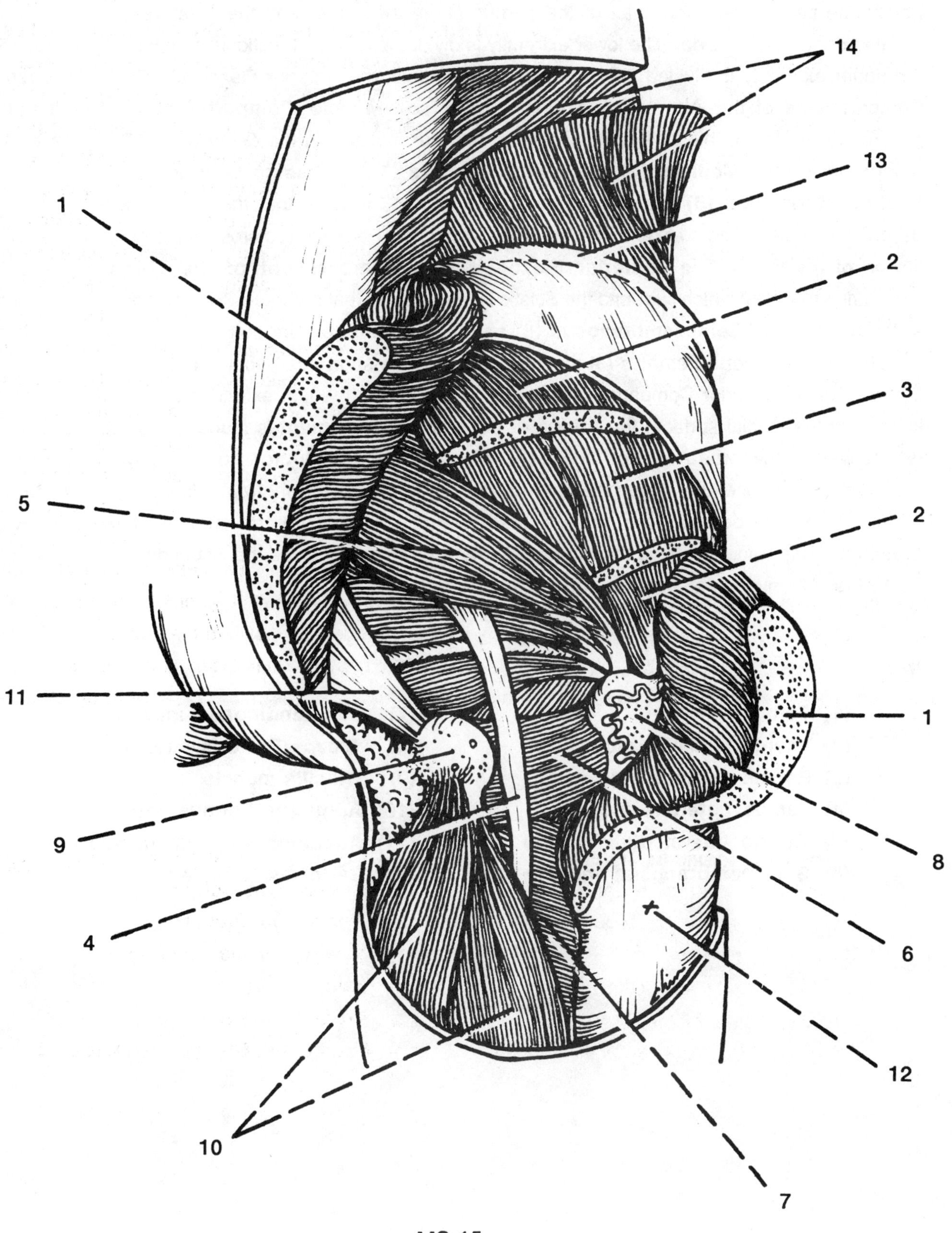

MS-15

Cross Section of the Right Thigh

(Inferior View)

This is a view of the Thigh viewed in cross section from below. Appreciate the orientation of this transverse section, with the shaft of the **Femur (1)** in the center and the great vessels of the thigh situated on the medial side. The lower extremity is divided into three functional compartments, much like the upper extremity is divided into two such components, the Anterior Flexor and Posterior Extensor Compartments of the Arm. The **Anterior Extensor Muscle Compartment of the Thigh** is pictured at the top of this view and consists of the four muscles in the **Quadriceps Femoris** muscle group: the **Vastus Medialis (2)**, the **Rectus Femoris (3)**, the **Vastus Lateralis (4)**, and the deep **Vastus Intermedius (5)** shown adjacent to the Femur. Realize that these four muscles all insert beyond the knee and act to extend the leg, while the Rectus Femoris also crosses the hip joint to assist in flexion of the thigh. The **Posterior Flexor Muscle Compartment of the Thigh** contains the "Hamstring muscles" which surround the Sciatic Nerve. Recall that the three "Hamstring muscles" consist of the more medial **Semimembranosus (6)** muscle, the **Semitendinosus (7)** muscle pictured in the middle, and the **Biceps Femoris (8)** muscle containing the short head above and the larger long head below. These Posterior Compartment muscles all act to flex the leg, extend the thigh, and rotate the knee. On the medial surface, the **Adductor Muscle Compartment** is found to include the **Gracilis (9)**, as well as the three Adductor muscles, two of which, the **Adductor Longus (10)** and **Adductor Magnus (11)** are seen here, with the **Adductor Brevis** muscle absent from this view. The **Adductor Canal** is located between the **Vastus Medialis (2)** and the ventral surface of the **Adductor muscles**. This Canal contains the Femoral Artery and Vein as well as the Saphenous Nerve, with the **Sartorius (12)** muscle situated over the fascial roof of the Canal.

(1) Femur
(2) Vastus Medialis muscle
(3) Rectus Femoris muscle
(4) Vastus Lateralis muscle
(5) Vastus Intermedius muscle
(6) Semimembranosus muscle
(7) Semitendinosus muscle
(8) Biceps Femoris muscle
(9) Gracilis muscle
(10) Adductor Longus muscle
(11) Adductor Magnus muscle
(12) Sartorius muscle

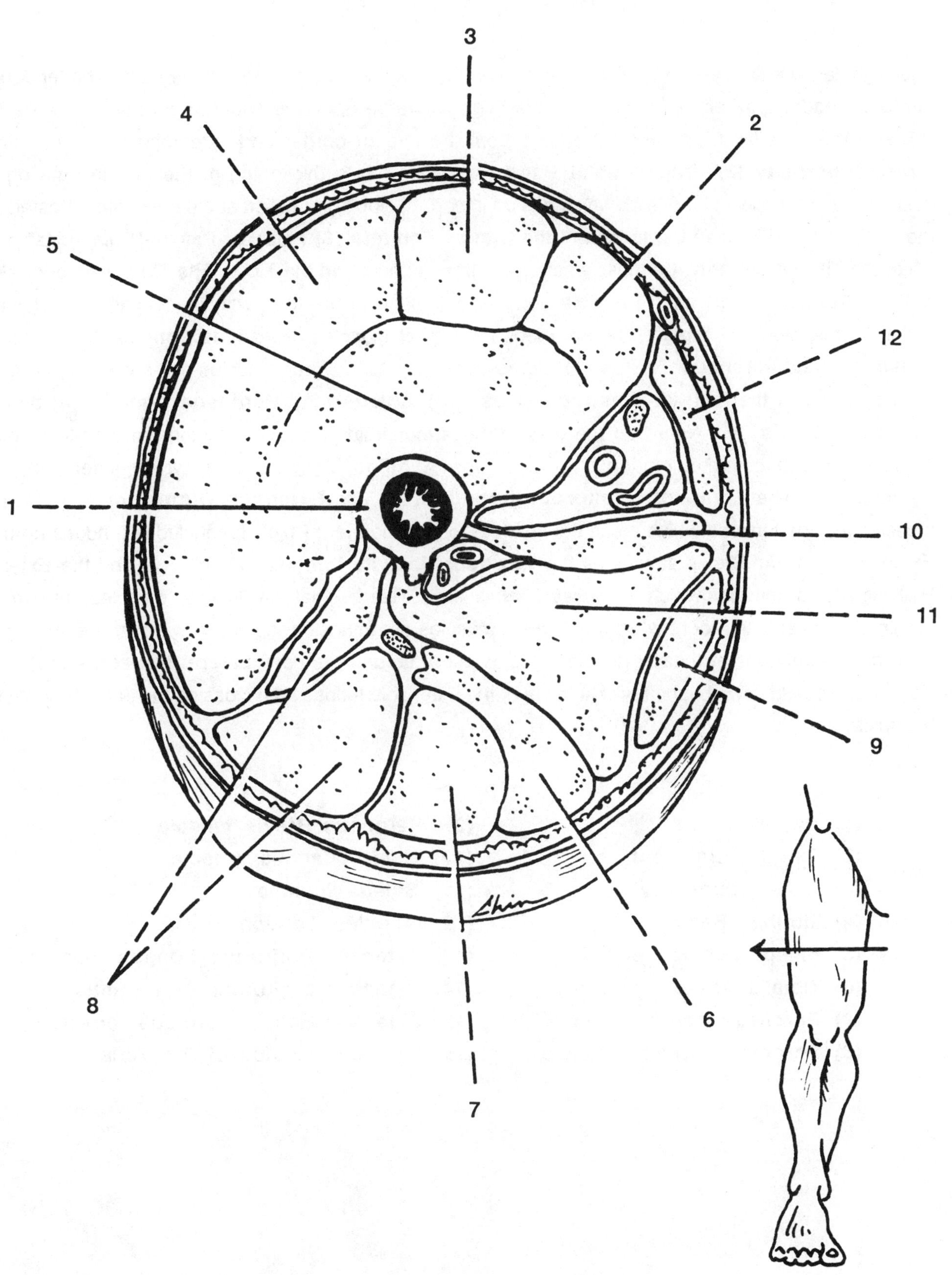
3
4
2
5
12
1
10
11
9
8
6
7
Chin

Superficial Leg Muscles

(Lateral and Anterior Views, Right side)

The Right leg is shown with a Lateral view seen on the left, the Anterior view on the right. The tendons of the four Quadriceps Femoris muscles of the thigh converge upon the superior surface of the **Patella (1)**, with the **Patellar Ligament (2)** arising from the inferior border and extending to the prominent **Tibial Tuberosity (3)**. The **Iliotibial Band (4)** or Tract is a thickening of the Fascia Lata on the posterior-lateral aspect of the thigh which has an important stabilizing action at the knee joint. Posterior to the Iliotibial Band is seen the tendon of the **Biceps Femoris (5)** muscle. Painful "Shin splits" occur when the **Tibialis Anterior (6)** muscle separates from its origin on the Tibia. The Tibialis Anterior along with the **Peroneus Tertius (7)** muscle act in Dorsiflexion of the foot, with the Tibialis Anterior also acting in Inversion. All three of the Peroneus (or Fibula) muscles (**Peroneus Longus (8), Peroneus Brevis (9)**, and **Peroneus Tertius (7)**) act to evert the foot. Four muscles are shown which act in Plantarflexion of the foot: the **Gastrocnemius (10), Soleus (11), Peroneus Longus (8)** and the **Peroneus Brevis (9)**. Note that the massive Gastrocnemius muscle and the flat, fish-like Soleus muscle share a common tendon, known as the **Achilles Tendon (12)**, which inserts into the Calcaneus. The **Extensor Digitorum Longus (13)** and **Extensor Digitorum Brevis (14)** muscles include tendons which reach the last four toes. The great ("big") toe includes tendons from the Extensor Digitorum Brevis as well as the **Extensor Hallucis Longus (15)** muscle and the Extensor Hallucis Brevis muscles which is not seen in this view. Realize that the "longus" muscles are extrinsic (originating outside the foot), while the "brevis" muscles are intrinsic (originating within the foot). Notice also that the tendons of these extrinsic muscles curve around the bony **Lateral Malleolus (16)** of the Tibia and are held down by the resilient fibers of the Extensor Retinaculum crossing the anterior surface of the ankle.

(1) Patella
(2) Patellar Ligament
(3) Tibial Tuberosity
(4) Iliotibial Band
(5) Biceps Femoris muscle
(6) Tibialis Anterior muscle
(7) Peroneus Tertius muscle
(8) Peroneus Longus muscle
(9) Peroneus Brevis muscle
(10) Gastrocnemius muscle
(11) Soleus muscle
(12) Achilles Tendon
(13) Extensor Digitorum Longus muscle
(14) Extensor Digitorum Brevis muscle
(15) Extensor Hallucis Longus muscle
(16) Lateral Malleolus of the Tibia

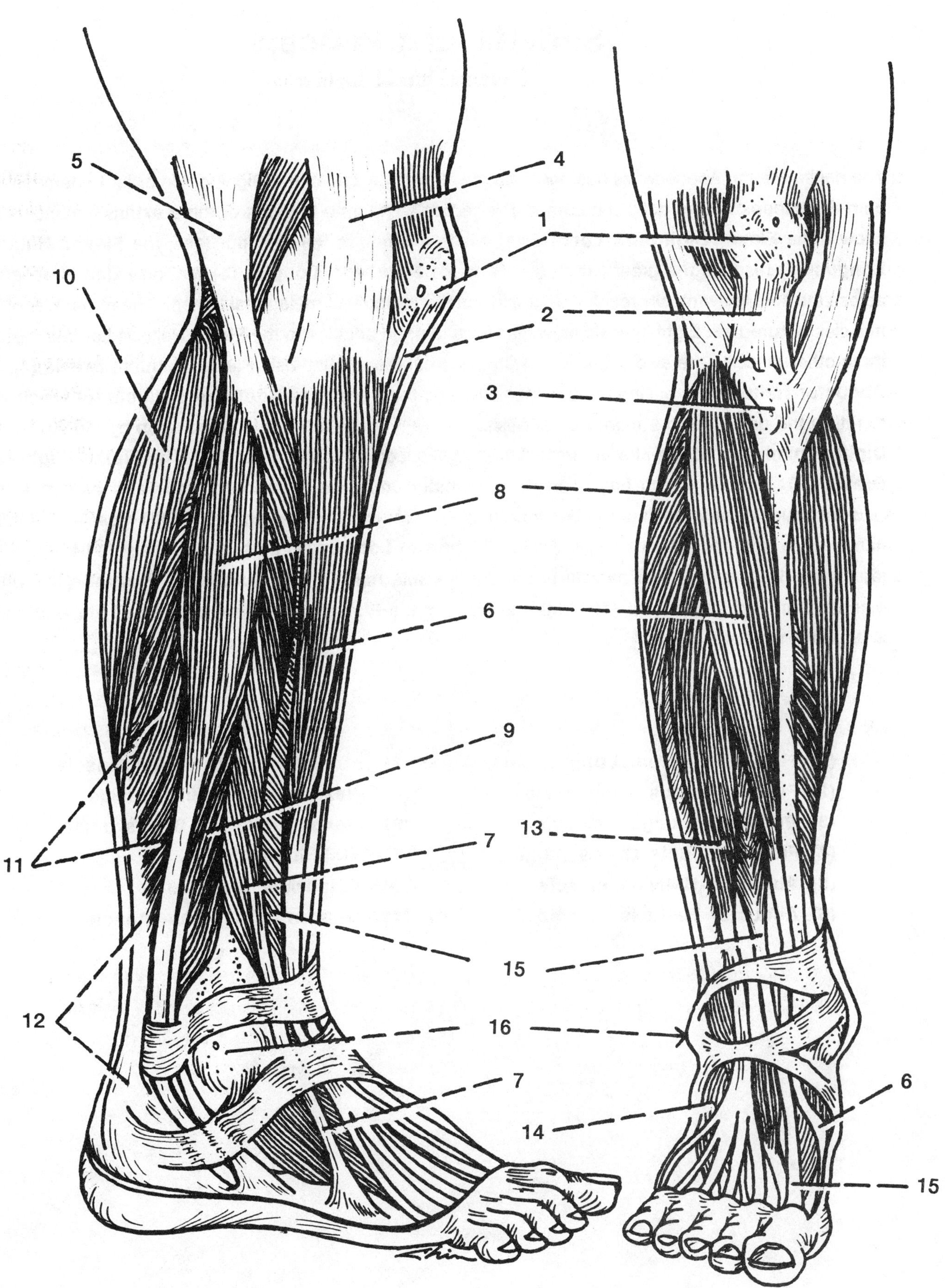

MS-17

Superficial and Deep Plantar Muscles

(Right side)

The dense Plantar Aponeurosis has been removed from the sole of the Right foot in order to demonstrate both the superficial and deep muscles in this region. The long tendons of three extrinsic muscles are shown: the **Flexor Digitorum Longus (1)** which extends to the last four toes, the **Flexor Hallucis Longus (2)** reaches the great ("big") toe, and the **Peroneus Longus (3)** tendon extends obliquely across the sole to eventually reach the medial cuneiform and first metatarsal bones. The remainder of the muscles pictured are Eight intrinsic muscles of the foot. In addition to the Flexor Hallucis Longus muscle, three other muscles are also associated with the great toe: the intrinsic **Flexor Hallucis Brevis (4)**, the **Abductor Hallucis (5)** pulling the toe medially, and the transverse **Adductor Hallucis (6)** which acts to return the great toe to a more lateral position. The fifth ('little") toe is acted upon by the **Abductor Digiti Minimi (7)** and the **Flexor Digiti Minimi (8)** much as in the great toe. The **Flexor Digitorum Brevis (9)** muscle (which has been cut) originates from the **Calcaneus (10)** and produces four tendons which pass to each of the last four toes. The four worm-like **Lumbrical (11)** muscles, as in the hand, arise from the tendons of the **Flexor Digitorum Longus (1)** muscle and assist in flexion of the toes. The last intrinsic foot muscle is the square sole muscle, the **Quadratus Plantae (12)** which extends from the **Calcaneus (10)** to the tendons of the **Flexor Digitorum Longus (1)** to also assist in flexion of the last four toes.

(1) Flexor Digitorum Longus muscle
(2) Flexor Hallicus Longus muscle
(3) Peroneus Longus muscle
(4) Flexor Hallucis Brevis muscle
(5) Abductor Hallucis muscle
(6) Adductor Hallucis muscle
(7) Abductor Digiti Minimi muscle
(8) Flexor Digiti Minimi muscle
(9) Flexor Digitorum Brevis muscle
(10) Calcaneus
(11) Lumbrical muscles
(12) Quadratus Plantae muscle

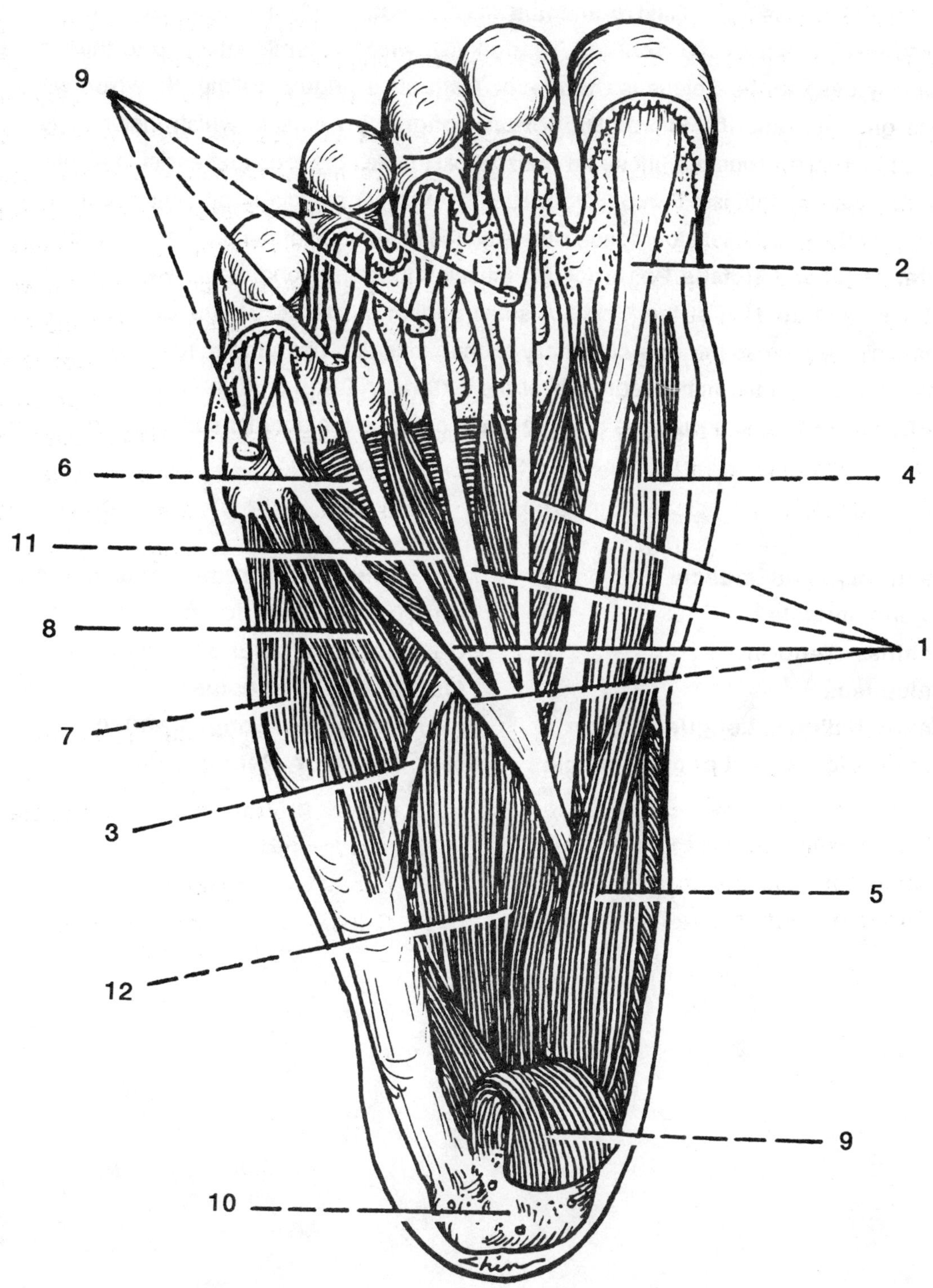
9
2
6
4
11
8
1
7
3
5
12
9
10
Chin

Leg and Plantar Muscles

(Right side)

The superficial muscles of the Right distal leg and sole of the foot are pictured to demonstrate the orientation of their tendons. The **Gastrocnemius (1)** and **Soleus (2)** muscles merge into a common tendon of insertion, known as the **Achilles Tendon (3)**, which is firmly attached to the **Calcaneus (4)**. Immediately deep to the Soleus is the **Flexor Hallucis Longus (5)** muscle which will send its tendon to the great toe and the **Flexor Digitorum Longus (6)** muscle which includes tendons to each of the last four toes. Together, these two muscles act to flex the toes and assist in plantarflexion of the foot. Notice that the intrinsic **Flexor Digitorum Brevis (7)** muscle is present deeper in the sole and also sends tendons to the last four toes. Note also the orientation of the tendons associated with the **Tibialis Anterior (8)** and **Tibialis Posterior (9)** muscles as these tendons curve around either side of the **Medial Malleolus (10)** of the Tibia. A small portion of the **Peroneus Longus (11)** tendon may be seen which will cross the sole diagonally to then inserting upon the medial cuneiform and first metatarsal bones. One of the many segments of the **Extensor Retinaculum (12)** is shown extending out from the **Calcaneus (4)** to secure the many tendons crossing the anterior surface of the ankle.

(1) Gastrocnemius muscle
(2) Soleus muscle
(3) Achilles Tendon
(4) Calcaneus
(5) Flexor Hallucis Longus muscle
(6) Flexor Digitorum Longus muscle
(7) Flexor Digitorum Brevis muscle
(8) Tibialis Anterior muscle
(9) Tibialis Posterior muscle
(10) Medial Malleolus
(11) Peroneus Longus muscle
(12) Extensor Retinaculum

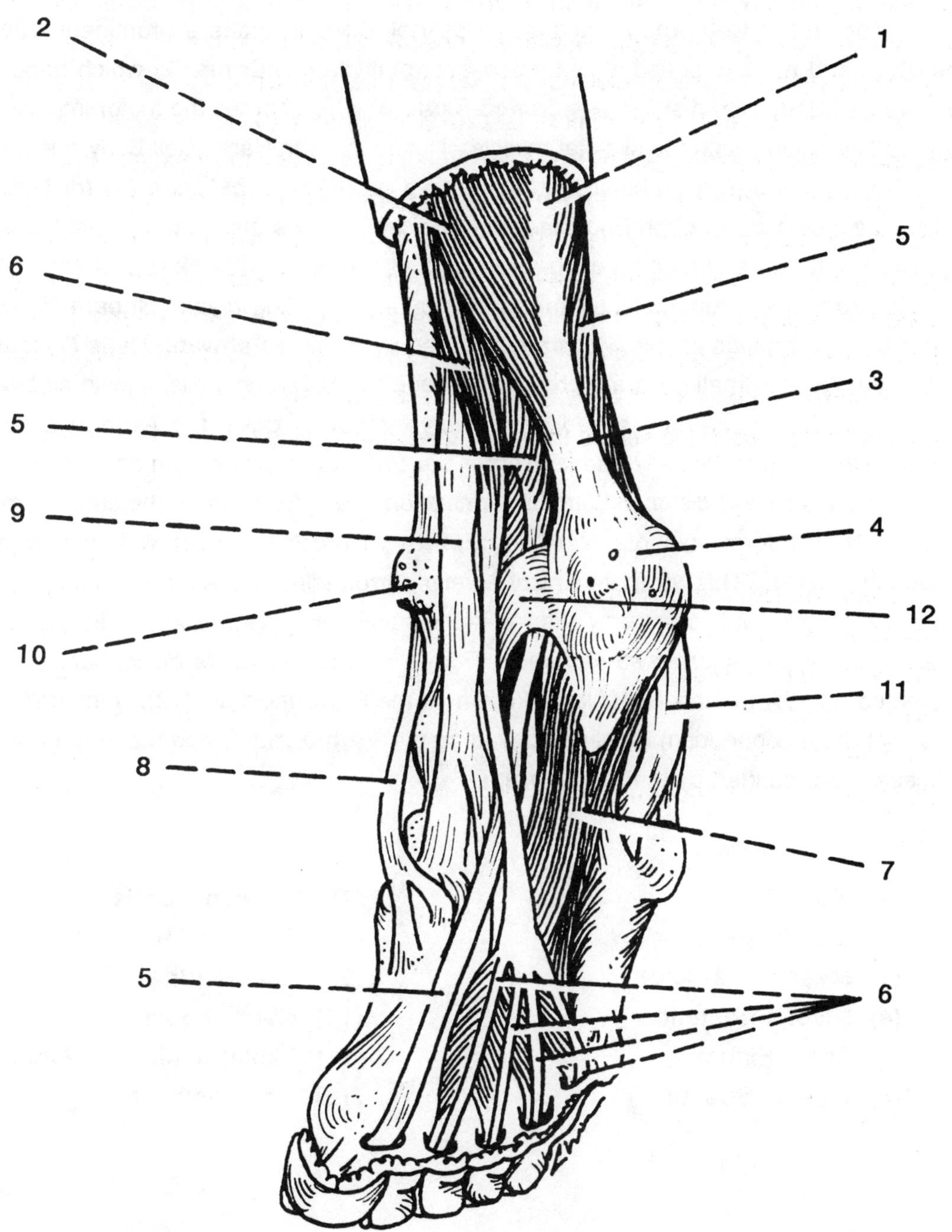

MS-19

Histology of a Peripheral Nerve Cell

A typical Motor neuron of the Peripheral Nervous System is pictured here. A neuron of this type originates in the Spinal Cord to become one component of a spinal nerve which will supply a skeletal muscle. The **Nerve Cell Body** found in the **Spinal Cord** includes a prominent **Nucleus (1)** with a **Nucleolus**, and two categories of cell processes: multiple **Dendrites (2)** which bring information into the Nerve Cell Body, and a single **Myelinated Axon (3)** which carries the motor impulses away from the neuron to eventually reach the skeletal muscle. Notice that the Nerve Cell Body and all of the Dendritic branches are filled with a granular material known as **Nissl Substance (4)** (or Nissl bodies) which consists of concentrated Rough Endoplasmic Reticulum. Notice also that no Nissl Substance is present in the Axon, with this clear region often referred to as the **Axon Hillock (5)**. A short distance from the Nerve Cell body, the Axon becomes surrounded by an insulating **Myelin Sheath (6)** which represents the concentric wrappings of lipid which are secreted by multiple **Schwann Cells (7)** found along the full length of the Axon. Small gaps are present between the Schwann Cells, known as **Nodes of Ranvier (8)**, which act to increase the speed of the nerve impulses down the Axon in a mechanism called **Saltatory Conduction** (literally, the leaping or dancing transmission from one Node to the next). The Myelination ends a short distance from the termination of the Axon where the small terminal branches or **Telodendria (9)** form multiple **Neuromuscular Junctions (10)** with the adjacent **Skeletal Muscle Fiber (11)**. This arrangement of Myelin surrounding the Axon is usually found in the large Multipolar Neurons, yet is commonly missing in the smaller neurons which are designated Unmyelinated Nerves. Finally, the Axon is surrounded by three concentric layers of connective tissue which are represented here by the innermost layer known as the **Endoneurium (12)**. Although not evident here, realize that this Endoneurium in then encased by the **Perineurium**, and the entire bundle of nerve cell processes is surrounded by the **Epineurium**.

(1) Nucleus
(2) Dendrites
(3) Myelinated Axon
(4) Nissl Substance
(5) Axon Hillock
(6) Myelin Sheath
(7) Schwann Cells
(8) Nodes of Ranvier
(9) Telodendria
(10) Neuromuscular Junctions
(11) Skeletal Muscle Fiber
(12) Endoneurium

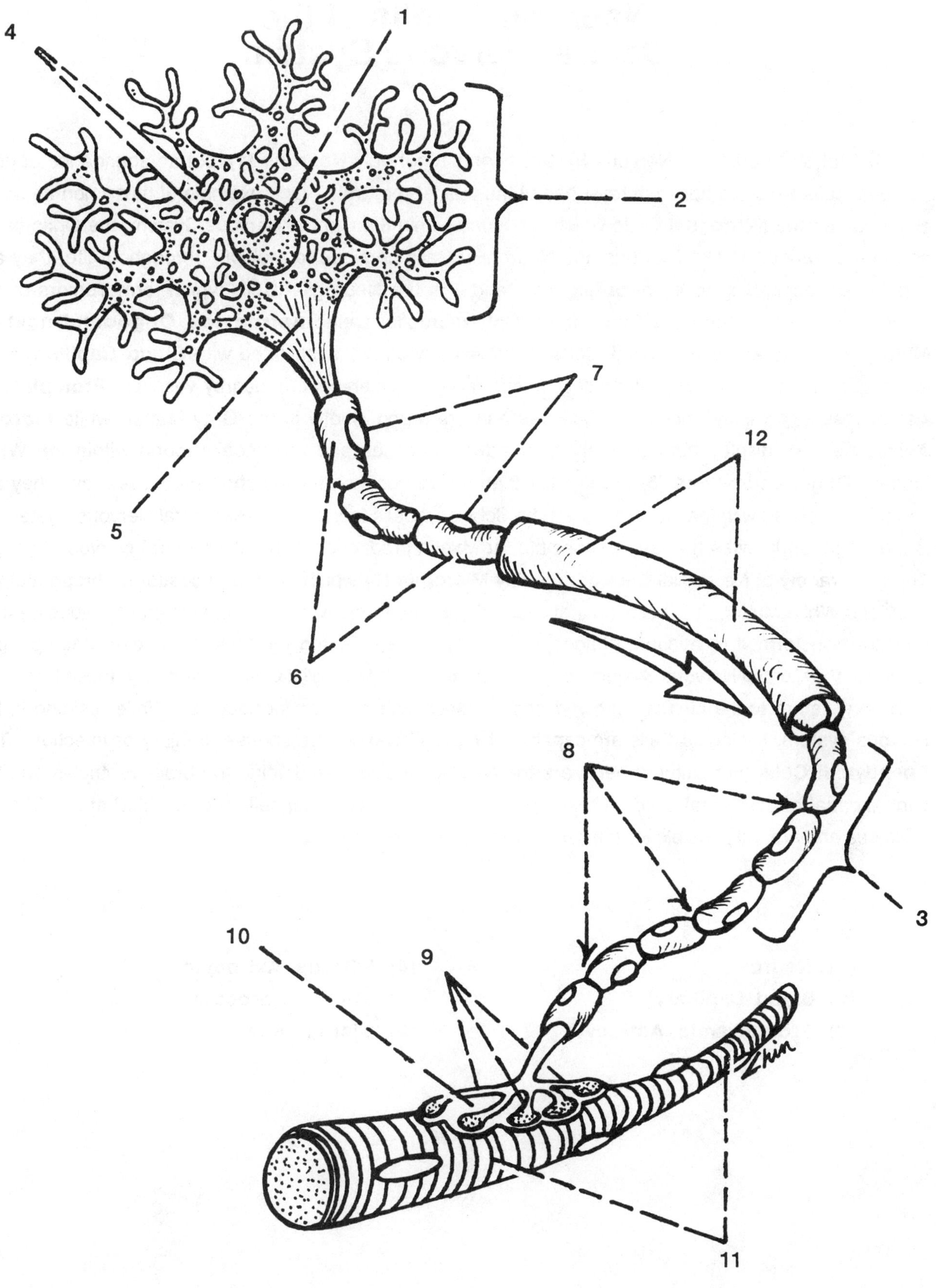

NS-1

Neuroglial Cells of the Central Nervous System

The Central and Peripheral Nervous Systems contain not only **Neurons (1)**, but an abundance of non-nervous cells which comprise at least half of the bulk of the Brain. One example of these non-neuronal elements are the **Neuroglial Cells** which are found within the Central Nervous System, the Optic nerve and in the Retina. Unlike Neurons, the Neuroglial Cells continue to multiply throughout life, they are capable of responding to injury or infection, and it is the Neuroglial Cells which form Tumors of the Nervous System. There are four varieties of Neuroglial Cells: **Astrocytes, Oligodendrocytes, Microglial Cells** and **Ependymal Cells**. The Astrocytes are associated with **Blood Capillaries (2)** and display multiple foot-like cell processes extending between several nearby vessels. **Protoplasmic Astrocytes (3)** display thick cell processes and are found chiefly in the Gray Matter, while **Fibrous Astrocytes (4)** display numerous short, slender processes and are usually found within the White Matter. **Oligodendrocytes (5)** are smaller than the Astrocytes, having short processes, and they are usually associated with the Neurons. Like the Schwann Cells found in the Peripheral Nervous System, it is the Oligodendrocytes that are responsible for Myelin production within the Central Nervous System. The third variety of Neuroglial Cells are the tiny **Microglia (6)** which are found scattered throughout the Gray and White Matter. It is believed that circulating Blood Monocytes enter the Central Nervous System and are transformed to become phagocytic Macrophages. During the time that these Macrophages remain in the Central Nervous System, they are referred to as Microglial Cells. Eventually, these Microglial Cells may return to the circulating blood and will once again act as Monocytes. While residing in the nervous system, Microglial Cells are capable of rapidly dividing in response to injury or infection. The **Ependymal Cells** (not pictured here) are the Neuroglial cells found lining the brain ventricles and the central canal of the Spinal Cord. These specialized ciliated cuboidal cells are modified at the Choroid Plexuses and are responsible for the secretion of Cerebrospinal fluid.

(1) Neuron
(2) Blood Capillary
(3) Protoplasmic Astrocyte
(4) Fibrous Astrocyte
(5) Oligodendrocyte
(6) Microglia

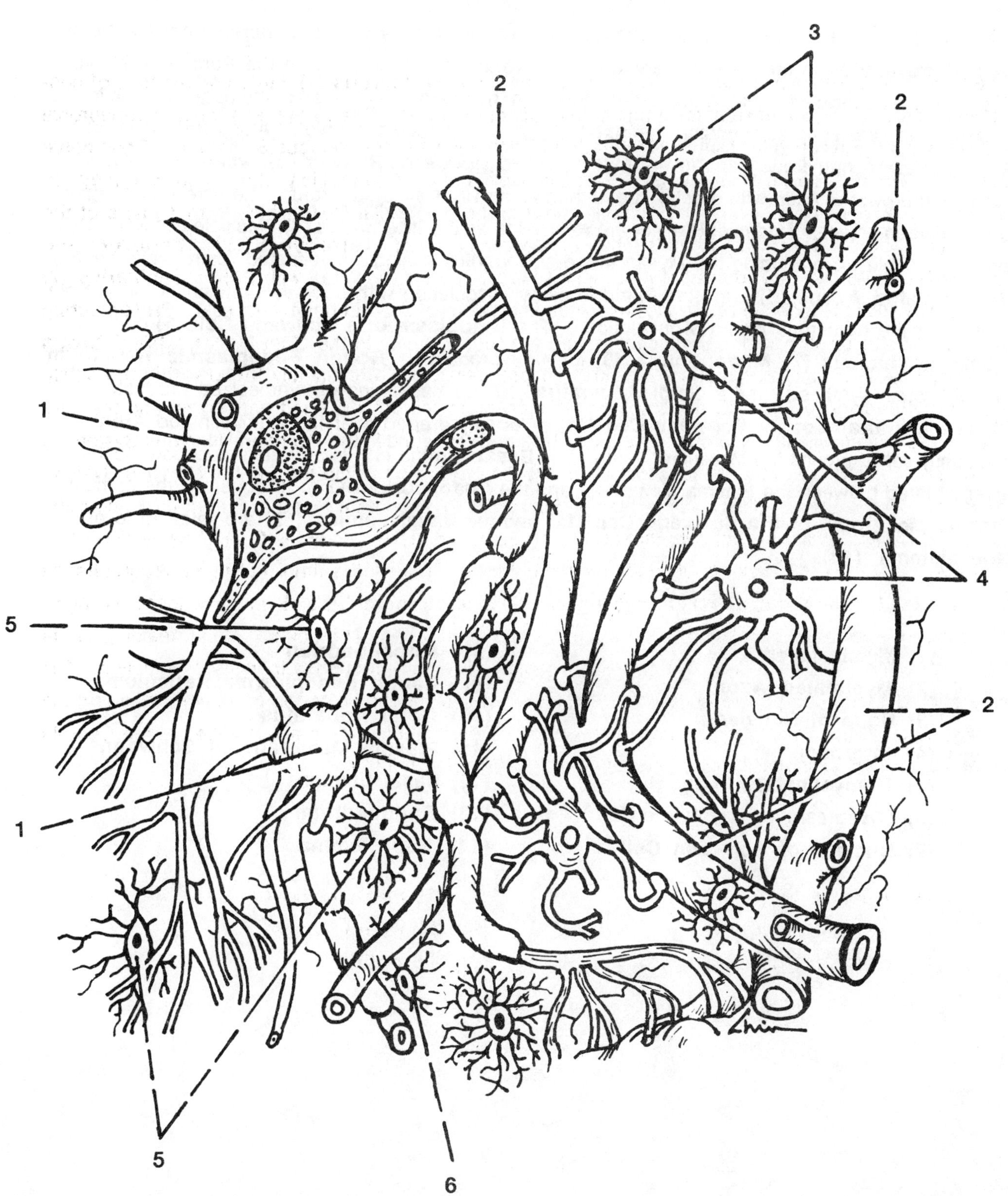
3
2
2
1
4
5
2
1
5
6

Myelinated and Unmyelinated Nerves

A **Myelinated** peripheral nerve is illustrated above, an **Unmyelinated** nerve appears below. **Oligodendrocytes** in the Central Nervous System and **Schwann Cells** in the Peripheral Nervous System are responsible for the production of Myelin. In the example at the top of the page, several layers of **Myelin (1)** are seen surrounding the **Axon (2)** in a tight circular fashion. Along the extent of the Axon, there are numerous gaps in the Myelin, called **Nodes of Ranvier (3)**, where the impulse is conducted down the Axon with greater speed than along Unmyelinated Axons. The myelinated regions seen between adjacent Nodes of Ranvier are referred to as the **Internodes (4)**. Multiple Schwann Cells are responsible for producing the Myelin found surrounding a single Myelinated Axon, while several **Unmyelinated Axons (5)** are often found in close association with a single Schwann Cell, yet not containing the numerous layers of insulation. The **Cytoplasm of a Schwann Cell (6)** contains a prominent **Nucleus (7)**, **Mitochondria (8)** for ATP production, **Rough Endoplasmic Reticulum (9)** for protein synthesis, and a **Golgi Apparatus (10)** for packaging of synthesized products. The **Plasma Membrane of the Schwann Cell (11)** is loosely draped over the Unmyelinated Axons, yet it is tightly coiled around the Myelinated Axon. The **Basal Lamina (12)** is a protein-mucopolysaccharide layer situated between the Plasma Membrane and the **Endoneurium (13)**. Frequently, the combined **Plasma Membrane of the Schwann Cell (11)** and the **Basal Lamina (12)** are referred to as the **Neurolemma (14)**.

(1) Myelin layer
(2) Myelinated Axon
(3) Node of Ranvier
(4) Internode
(5) Unmyelinated Axons
(6) Cytoplasm of Schwann Cell
(7) Nucleus of Schwann Cell
(8) Mitochondrion
(9) Rough Endoplasmic Reticulum
(10) Golgi Apparatus
(11) Plasma Membrane of Schwann
(12) Basal Lamina
(13) Endoneurium
(14) Neurolemma

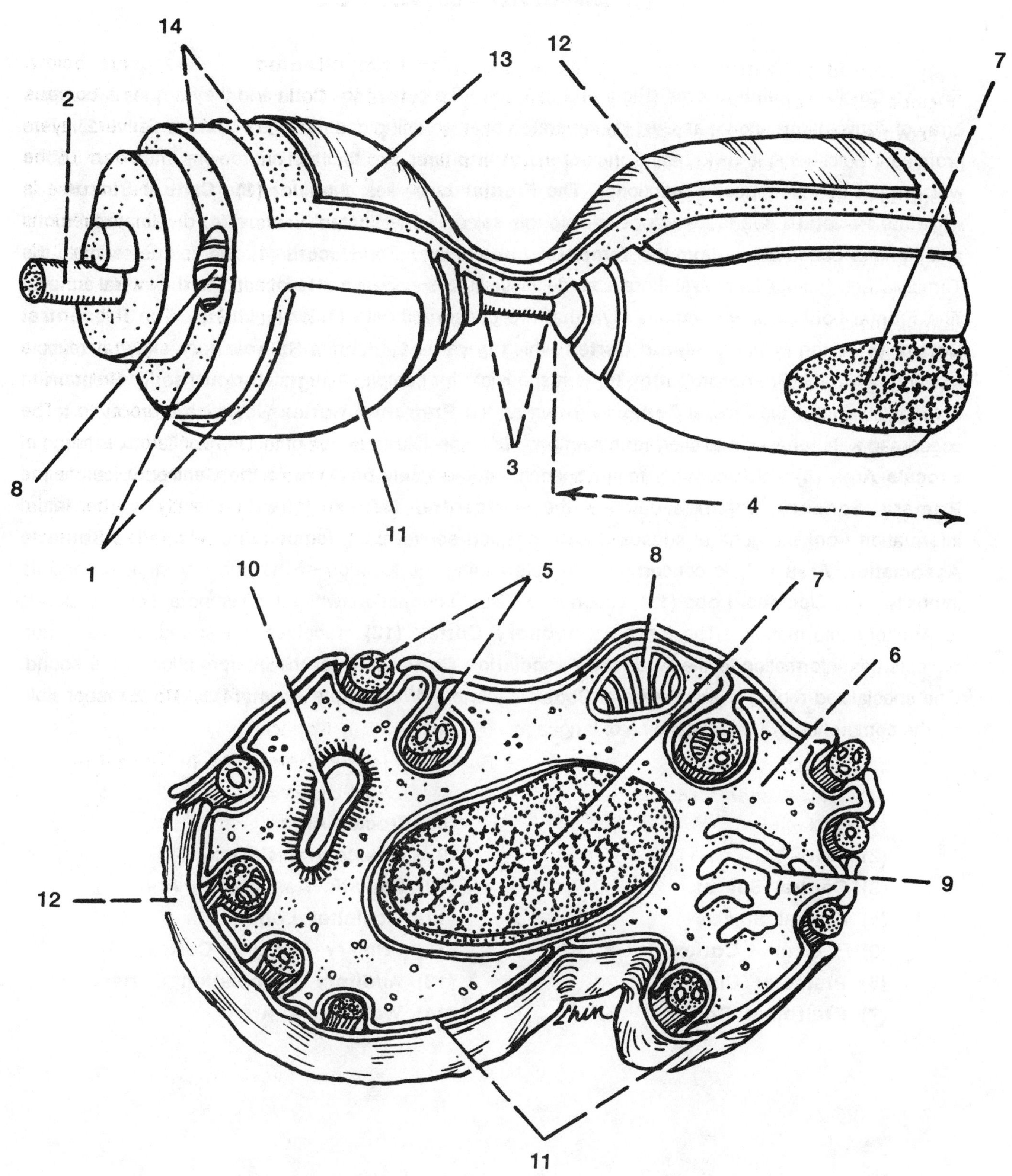
14
12
13
7
2
8
3
4
11
1
10
5
8
7
6
12
9
11

Cerebral Cortex

(Lateral View, Left side)

The two **Cerebral Hemispheres** (Right and Left) form the bulk of the brain and they display a complex array of convolutions known as **Gyri (1)** separated by intervening depressions known as **Sulci (2)**. Two prominent Sulci serve to divide the Cortex of the hemispheres into four distinct regions known as **Lobes** which each conduct specific functions. The **Frontal Lobe** lies anterior to the **Central Sulcus (3)**, while the **Parietal Lobe** is found posterior to this sulcus. The **Lateral Sulcus (4)** divides the Frontal and Parietal Lobes above from the **Temporal Lobe** below. The **Occipital Lobe** lies posterior to the Temporal and Parietal lobes, yet there is no definitive sulcus evident in this location on the lateral surface. The Frontal Lobe contains various Gyri that are concerned with Motor functions. The **Precentral Cortex (5)** is the **Primary Motor Cortex** which governs the control of voluntary skeletal muscle function, while the **Premotor Cortex (6)** is responsible for complex integrated motions. The remaining anterior portion of the Frontal Cortex is known as the **Prefrontal Cortex (7)** which is believed to be concerned with behavior and short-term memory. One specialized region of the Frontal Cortex is found in **Broca's Area (8)** which controls motor speech. Immediately posterior to the Central Sulcus is the **Primary Sensory Cortex** situated in the **Postcentral Cortex (9)** which receives the initial information from the general senses (touch, position sense, pain, temperature, etc.). The **Somatic Association Area (10)** is concerned with determining the location of the sensory stimulus and its intensity. The **Occiptial Lobe (11)** responds to Visual information, while the Temporal Lobe responds to Auditory information. The **Primary Auditory Cortex (12)** receives the sound impulses and projects this information to the **Auditory Association Cortex (13)** for the interpretation of the sound. One specialized region of the Temporal Cortex is found in **Wernicke's Area (14)** which is responsible for the comprehensive understanding of language.

(1) Gyri
(2) Sulci
(3) Central Sulcus
(4) Lateral Sulcus
(5) Precentral Cortex
(6) Premotor Cortex
(7) Prefrontal Cortex
(8) Broca's Area
(9) Postcentral Cortex
(10) Somatic Association Area
(11) Occipital Lobe
(12) Primary Auditory Cortex
(13) Auditory Association Cortex
(14) Wernicke's Area

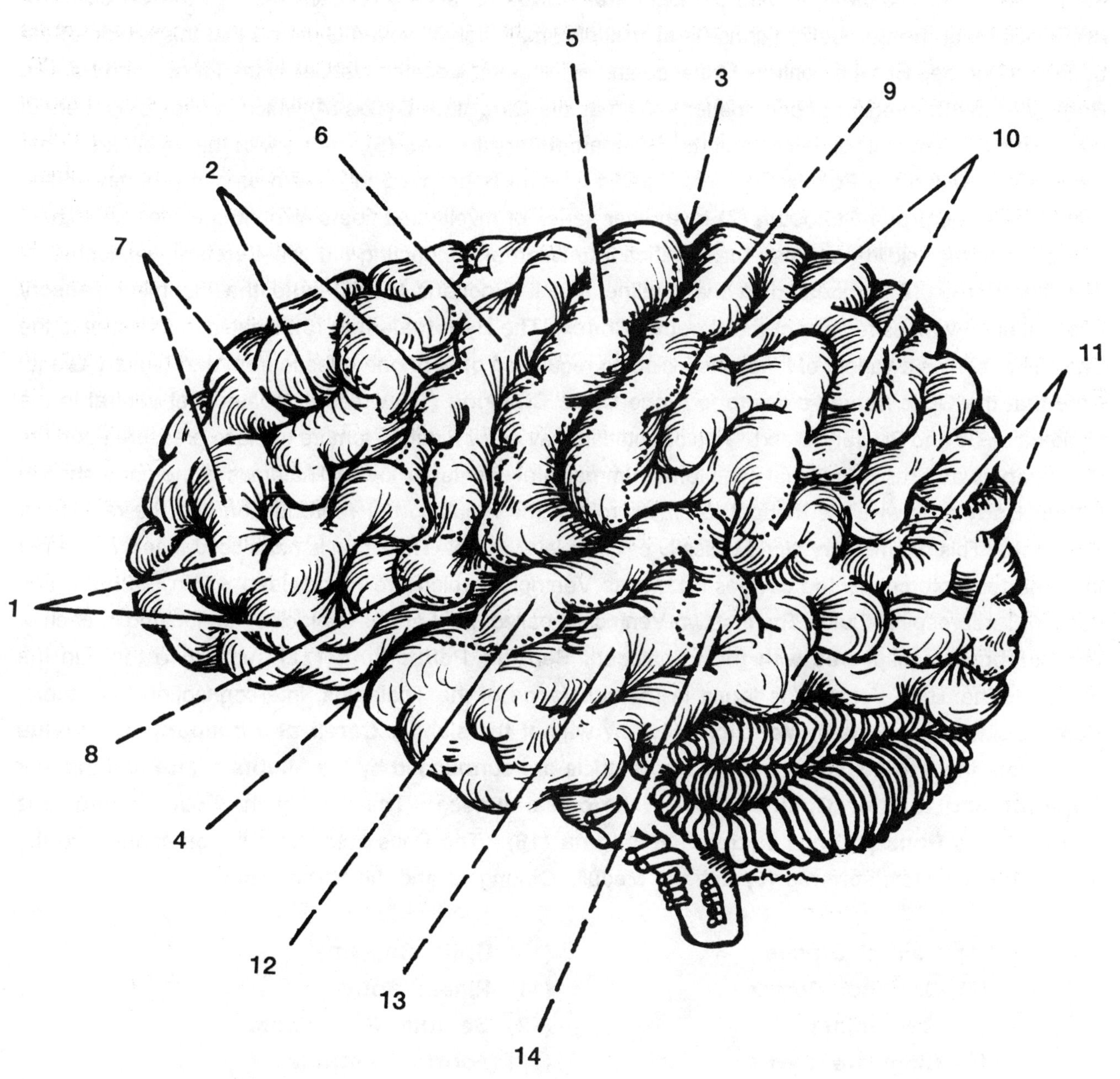
5
3
9
6
10
2
7
11
1
8
4
12
13
14

Midsagittal Brain

(Midsagittal Section)

This midsagittal view of the Right Cerebral Hemisphere reveals that the **Central Sulcus (1)** continues from the lateral surface of the brain onto this surface, thereby distinguishing the Frontal Cortex to the left, the Parietal Cortex to the right, and the **Occipital Cortex (2)** situated posterior above the **Cerebellum (3)**. Notice that the Cerebellum consists of multiple small, tightly woven Gyri and that this region of the brain is concerned with the control of balance and equilibrium. Located just below the Frontal and Parietal Cortices is another region of gray matter known as the **Cingulate Gyrus (4)** which is one component of the Limbic System that also includes the two **Mammillary Bodies (5)**, along with the myelinated fiber tracts found within the **Fornix (6)**. This Limbic System is believed to govern emotion, behavior and memory. The **Corpus Callosum (7)** is another series of myelinated fibers forming a white matter tract that crosses the midline and has the significant function of connecting the two Cerebral Hemispheres. The **Thalamus (8)** is located in the midline and it functions to distribute the incoming sensory information to various regions of the Cerebral Cortex. The **Hypothalamus (9)**, located just ventral to the Thalamus, serves a variety of functions including regulation of hormone release from the Pituitary Gland. Note that the Optic Nerves cross to form the **Optic Chiasma (10)** which is located just ventral to the Hypothalamus and Pituitary Gland. This orientation may allow Pituitary tumors to place pressure upon the Optic nerves and result in visual symptoms. Immediately posterior to the Thalamus is the cone-shaped **Pineal Body (11)** which is thought to play a role in the regulation of reproduction and the release of Melatonin. This rudimentary gland typically calcifies later in life. The Brain is not a solid mass, but rather includes four interconnected cavities known as Ventricles which are found deep within the brain and contain Cerebrospinal Fluid. The first two Ventricles, called the Lateral Ventricles, are located in each of the Cerebral Hemispheres with the translucent **Septum Pellucidum (12)** separating them in the midline. The Third Ventricle is found in the substance of the Thalamus, interconnecting the Lateral Ventricles with the **Fourth Ventricle (13)** by way of the slender **Cerebral Aqueduct of Sylvius (14)**. Note that the Aqueduct and Fourth Ventricle are surrounded by the **Midbrain (15)** and the four **Superior and Inferior Colliculi (16)** on the dorsal surface. The floor of the Fourth Ventricle is formed by the **Pons (17)** and **Medulla Oblongata (18)**. The Pons acts like a bridge connecting the two **Cerebellar Hemispheres (3)** with the Medulla Oblongata and the Spinal Cord.

(1) Central Sulcus
(2) Occipital Cortex
(3) Cerebellum
(4) Cingulate Gyrus
(5) Mammillary Body
(6) Fornix
(7) Corpus Callosum
(8) Thalamus
(9) Hypothalamus
(10) Optic Chiasma
(11) Pineal Body
(12) Septum Pellucidum
(13) Fourth Ventricle
(14) Cerebral Aqueduct of Sylvius
(15) Midbrain
(16) Superior & Inferior Colliculi
(17) Pons
(18) Medulla Oblongata

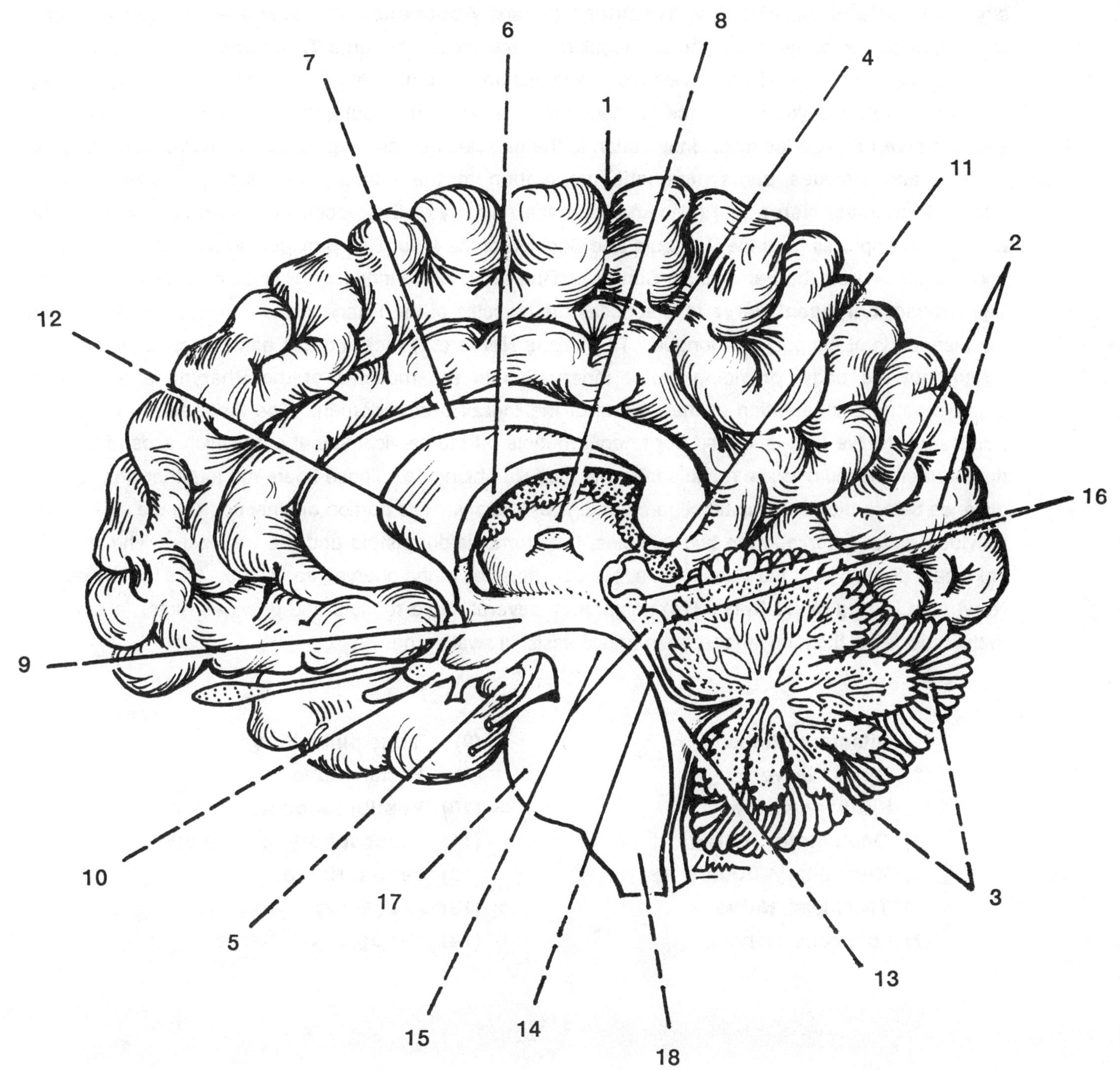

8
6
4
7
1
11
2
12
16
9
10
17
5
3
13
15
14
18

Cranial Nerves

The Twelve pairs of Cranial Nerves can best be visualized from the Ventral surface of the brain. The **Olfactory Nerves** convey smell sensations from receptors high in the nasal cavity which synapse at the dilated ends of the Olfactory nerves known as the **Olfactory Bulbs (1)**. The **Optic Nerve (2)** carries visual information from the Retina and is seen to partially cross at the Optic Chiasma located just anterior to the **Pituitary Gland (3)**. The **Oculomotor Nerve (4)** emerges just lateral to the **Mammillary Bodies (5)** and provides motor innervation to four of the six extraocular eye muscles as well as the smooth muscle of the Iris. The **Trochlear (6)** and **Abducens (7) Nerves** each provide motor innervation to one of the remaining extraocular eye muscles. The large **Trigeminal Nerve (8)** consists of three major branches which provide motor innervation to the muscles of mastication and carry sensory information from the skin of the face, mucous membranes of the mouth, the teeth and the cornea. The **Facial Nerve (9)** provides motor innervation to the muscles of facial expression, conveys taste from the tongue, and includes parasympathetic innervation to the salivary and lacrimal glands. The **Vestibulocochlear Nerve (10)** (also known as the Auditory or Stato-acoustic Nerve) has two divisions which carry impulses from the Vestibule and Cochlea of the Inner Ear, providing information for balance and equilibrium from the Semicircular Canals and the sense of Hearing from Hair Cells within the Cochlea. The **Glossopharyngeal Nerve (11)** supplies the muscles of the Pharynx and conveys sensation from the taste buds of the posterior tongue. The **Vagus Nerve (12)** received this name because it wanders throughout the body, providing motor innervation to the muscles of the Pharynx and carrying parasympathetic innervation to the organs of the thorax and abdomen. The **Accessory** (or Spinal Accessory) **Nerve (13)** is formed from nerve rootlets off the cervical spinal cord which ascend through the foramen magnum to join rootlets from the medulla oblongata. These fibers run together only a short distance before they once again separate into two divisions. The portion originating from the spinal cord provides motor innervation to two muscles, the sternocleidomastoid and the trapezius. The medulla oblongata division joins the Vagus nerve to supply muscles of the pharynx and larynx. Like the Accessory Nerve, the **Hypoglossal Nerve (14)** arises from several nerve rootlets and provides motor innervation to the muscles of the tongue for speech and to assist in swallowing.

(1) Olfactory Bulbs
(2) Optic Nerve
(3) Pituitary Gland
(4) Oculomotor Nerve
(5) Mammillary Bodies
(6) Trochlear Nerve
(7) Abducens Nerve
(8) Trigeminal Nerve
(9) Facial Nerve
(10) Vestibulocochlear Nerve
(11) Glossopharyngeal Nerve
(12) Vagus Nerve
(13) Accessory Nerve
(14) Hypoglossal Nerve

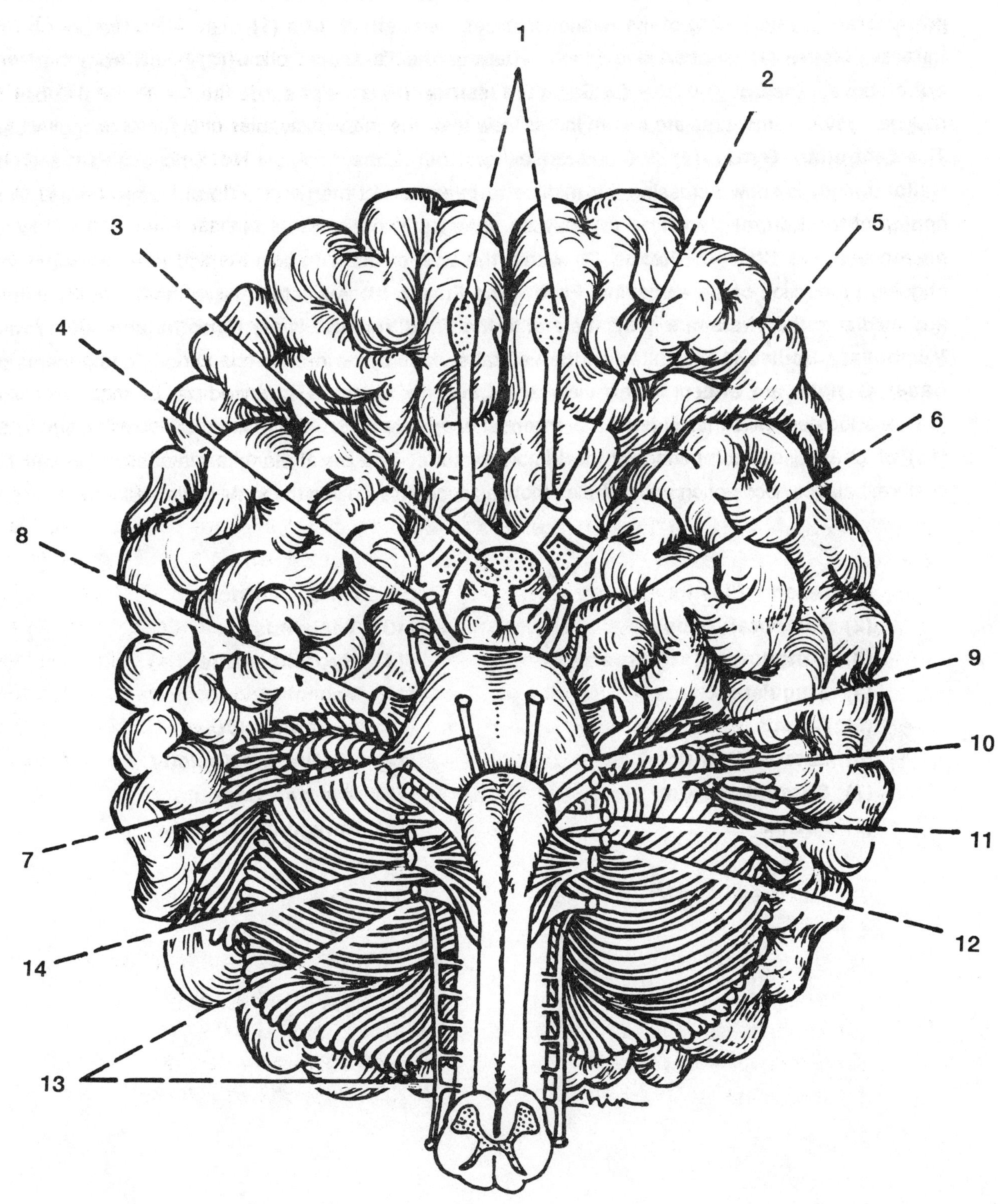
1
2
3
4
5
6
8
9
10
7
11
14
12
13

Coronal Brain Section

This Coronal or Frontal section permits visualization of various tracts as they run though the brain and provides an understanding of the relationship between certain brain regions. Note the position of the **Lateral Fissure (1)** on each side which separates the **Temporal Lobe (2)** below from the **Parietal Lobe** above. The **Longitudinal** (or Sagittal) **Fissure (3)** serves to divide the two Parietal Lobes in the midline. Several structures are shown in this view that were previously evident in the Midsaggital section. The **Cingulate Gyrus (4)** is found on either side of the Longitudinal Sulcus and the **Corpus Callosum (5)** is shown crossing the midline to interconnect the two Cerebral Hemispheres. A small portion of the **Lateral Ventricle (6)** may be seen as it extends through the Hemisphere. Note the presence of the **Choroid Plexus (7)** within the wall of the Ventricles which is responsible for the ongoing production of Cerebrospinal Fluid. The **Fornix (8)** is located above the **Third Ventricle (9)** and medial to the **Thalamus (10)**. Lateral to the Third Ventricle is the **Hypothalamus (11)** with the **Mammillary Bodies (12)** situated on the ventral surface of the brain. The various components of the **Basal Ganglia** are evident in this view: the **Caudate Nucleus (13)** and the **Lentiform Nucleus (14)** are located on either side of the **Internal Capsule (15)**, along with the **Amygdaloid Body (16)** which is also a component of the Limbic system. The Basal Ganglia plays an important role in posture, balance, locomotion and complex motions such as swinging the arms while walking.

(1) Lateral Fissure
(2) Temporal Lobe
(3) Longitudinal Fissure
(4) Cingulate Gyrus
(5) Corpus Callosum
(6) Lateral Ventricle
(7) Choroid Plexus
(8) Fornix
(9) Third Ventricle
(10) Thalamus
(11) Hypothalamus
(12) Mammillary Bodies
(13) Caudate Nucleus
(14) Lentiform Nucleus
(15) Internal Capsule
(16) Amygdaloid Body

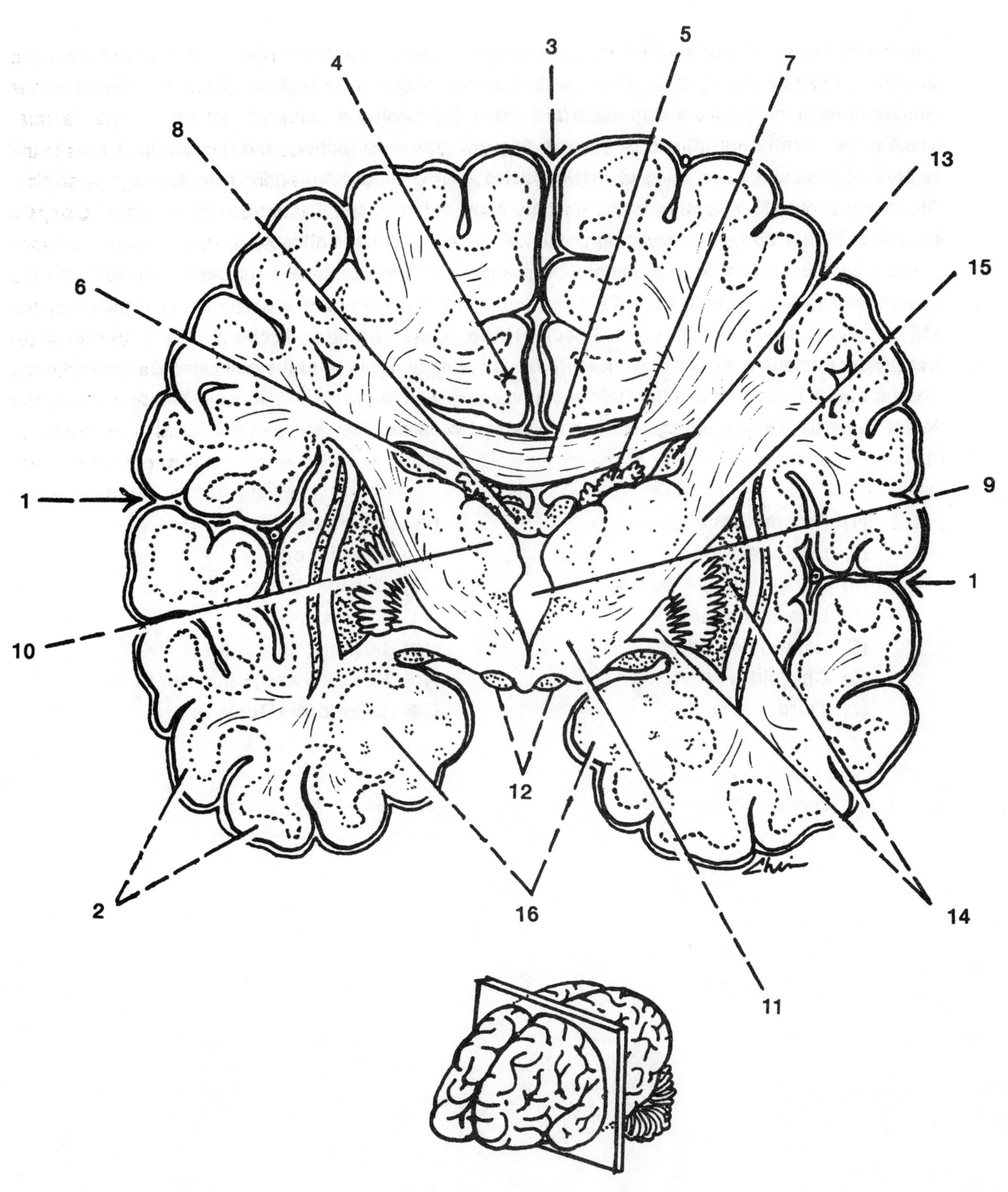
3
5
4
7
8
13
6
15
1
9
1
10
12
2
16
14
11

Horizontal Brain Section

This Transverse or Horizontal section of the brain passes through the **Frontal Lobe (1)** anteriorly and the **Occipital Lobe (2)** posteriorly, with the **Longitudinal Fissure (3)** separating the two Hemispheres in the midline and the **Lateral Fissure (4)** dividing the Frontal from the Temporal Lobes. The **Lateral Ventricles (5)** are seen extending into each Hemisphere and the **Choroid Plexus (6)** is present throughout the lining of the ventricles for the production of **Cerebrospinal Fluid**. The **Third Ventricle (7)** is located in the midline, surrounded by the **Thalamus (8)** which displays the cone-like **Pineal body (9)** extending posteriorly. The **Corpus Callosum (10)** is seen twice in this section as the anterior and posterior fibers arch from front to back in order to connect the two Hemispheres. Note that the **Fornix (11)** is seen in the midline of this section. The **Internal Capsule (12)** is shown separating the more medial **Head** of the **Caudate Nucleus (13)** from the lateral **Lentiform Nucleus (14)**. The Lentiform Nucleus is the name given to the combined Putamen and the Globus Pallidus. Realize that the Caudate curves backward and laterally, with the **Tail of the Caudate Nuclei** situated at the outer edge of the each Lateral Ventricle.

(1) Frontal Lobe
(2) Occipital Lobe
(3) Longitudinal Fissure
(4) Lateral Fissure
(5) Lateral Ventricle
(6) Choroid Plexus
(7) Third Ventricle
(8) Thalamus
(9) Pineal Body
(10) Corpus Callosum
(11) Fornix
(12) Internal Capsule
(13) Head of the Caudate Nucleus
(14) Lentiform Nucleus

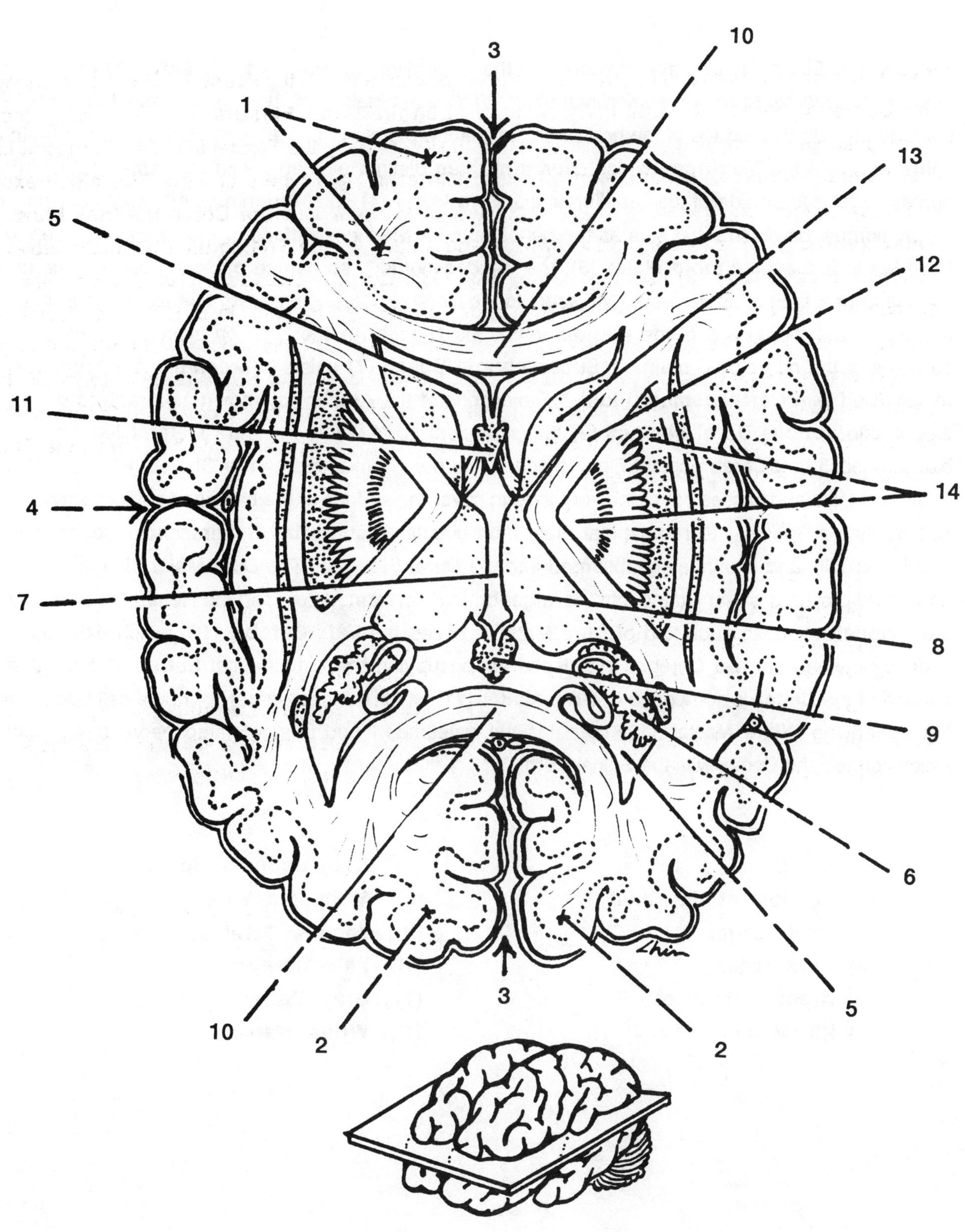
10
3
1
13
5
12
11
14
4
7
8
9
6
5
3
10
2
2
Chin

Meninges and Cerebrospinal Fluid

The durable **Scalp (1)** is firmly anchored to the underlying **Cranium of the Skull (2)** by the dense fibers of the **Periosteum** (or Pericranium) **(3)**. The combination of the Scalp, Periosteum and Skull serve to provide the surface of the brain with excellent protection. Covering the surface of the brain and spinal cord are the **Meninges** which consist of three distinct elements with the term "Mater" in their names. Lining the inside of the Cranium is the **Dura Mater (4)**, the "Tough mother", which consists of a tough, fibrous membrane that acts as the periosteum for the inner surface of the Cranium. Deep to the Dura Mater is the **Arachnoid Mater (5)**, the "Spider mother", so named because this thin membrane has the appearance of spider webs! Adjacent to the Cerebral Cortex and outer surface of the Spinal Cord is the delicate membrane of the **Pia Mater (6)**, the "Tender mother", which closely follows all of the contours of the convoluted brain. A **Subarachnoid Space (7)** exists between the Arachnoid and Pia mater, filled with **Cerebrospinal Fluid**. Projections of the Arachnoid, known as **Arachnoid Villi (8)**, extend into **Venous Dural Sinuses (9)** that represent splits within the Dura Mater, filled with venous blood. Recall that Cerebrospinal Fluid is constantly being produced by the Choroid Plexus within the brain ventricles and passes out of the ventricles to cover the surface of the brain and spinal cord within the Subarachnoid Space. Cerebrospinal Fluid then circulates out of the Subarachnoid Space, into the Arachnoid Villi, and is ultimately absorbed into the Dural Sinuses. The Dura Mater also includes long downward projections that extend into the larger brain Sulci located between the Hemispheres and above the Cerebellum. One such fold of Dura Mater is called the **Falx Cerebri (10)** which extends in the midline between the two Cerebral Hemispheres, occupying the Longitudinal Sulcus. Notice that the surface of the Cortex is composed of **Gray Matter (11)** which consists largely of nerve cell bodies, while the underlying **White Matter (12)** contains the myelinated and unmyelinated nerve cell processes which connect the Cortex with the Spinal Cord.

(1) Scalp
(2) Cranium of the Skull
(3) Periosteum
(4) Dura Mater
(5) Arachnoid Mater
(6) Pia Mater
(7) Subarachnoid Space
(8) Arachnoid Villi
(9) Venous Dural Sinus
(10) Falx Cerebri
(11) Gray Matter
(12) White Matter

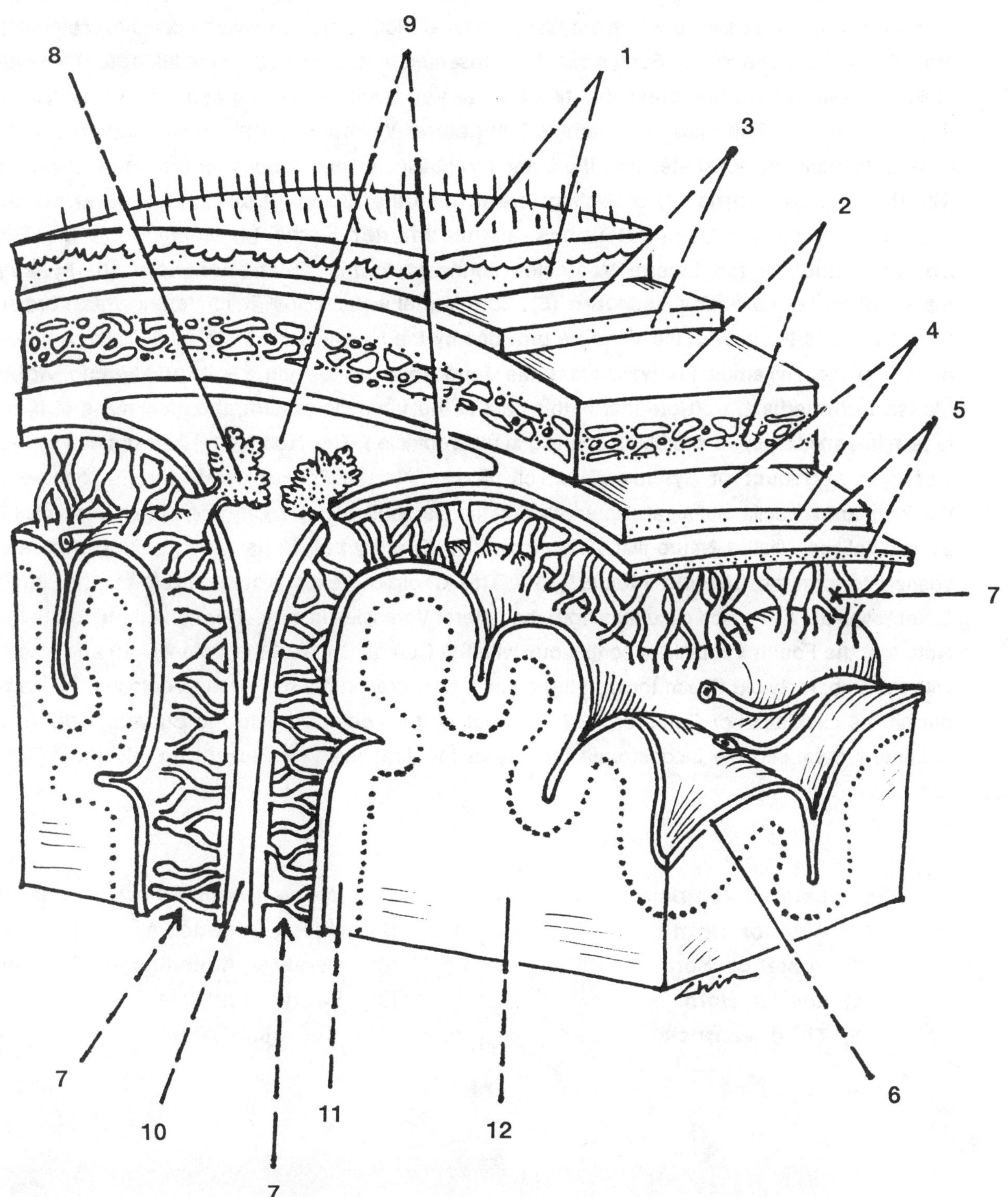
9
8
1
3
2
4
5
7
6
7
10
11
12
7

Brain Ventricles

The entire Nervous System develops from a hollow embryonic Neural Tube. As the Brain and Spinal Cord expand and complete their formation, the original canal within the Neural Tube persists to become the Brain Ventricles and Central Canal of the Spinal Cord. The fluid-filled Brain Ventricles are best visualized with the use of solid casts in which the Cerebrospinal Fluid is replaced with plastic and the surrounding brain tissue is then removed. Such a cast is represented here, viewed from the left side. There are a total of **Four Brain Ventricles** present: two Lateral Ventricles (Right and Left), with a single Third and Fourth Ventricle. This view of the entire Left **Lateral Ventricle (1)**, along with a portion of the Right Lateral Ventricle, demonstrates how the Lateral Ventricles extend throughout the Cerebral Hemispheres, with the **Anterior Horns (2)** projecting forward into the **Frontal Lobes**, the **Posterior Horns (3)** passing back into the **Occipital Lobes**, and the **Inferior Horns (4)** reaching into the **Temporal Lobes**. Each of the Lateral Ventricles connects to the **Third Ventricle (5)** by way of the **Interventricular Foramina of Monro (6)**. The lateral walls of the Third Ventricle are formed by the Thalamus, with the floor of the Ventricle provided by the Hypothalamus. In 70-80% of human brains, a portion of the Thalamus extends across the Third Ventricle to form the **Interthalamic Adhesion** or **Massa Intermedia (7)**. (Note that in this cast of the Ventricle system, the solid mass of tissue in the Massa Intermedia appears to be an opening in the Ventricle.) The Third Ventricle is drained by the slender **Cerebral Aqueduct of Sylvius (8)** which passes into the **Fourth Ventricle (9)** located between the Pons and Medulla Oblongata anteriorly and the Cerebellum posteriorly. The brain Ventricles are lined by specialized ciliated Neuroglial Cells known as **Ependymal Cells** which unite with nearby blood vessels to form the **Choroid Plexus (10)**. This Choroid Plexus is responsible for the production of Cerebrospinal Fluid which circulates from the Lateral Ventricles into the Third and then Fourth Ventricles. Note that the Fourth Ventricle is continuous with the Central Canal of the Spinal Cord and also connects with the Subarachnoid Space through three small apertures in the roof of the Ventricle. These openings permit the circulation of Cerebrospinal Fluid out of the Ventricles, into the Subarachnoid Space, and ultimately back into the bloodstream by way of the Arachnoid Villi extending into the Venous Dural Sinuses.

(1) Lateral Ventricle
(2) Anterior Horn
(3) Posterior Horn
(4) Inferior Horn
(5) Third Ventricle
(6) Interventricular Foramina of Monro
(7) Massa Intermedia
(8) Cerebral Aqueduct of Sylvius
(9) Fourth Ventricle
(10) Choroid Plexus

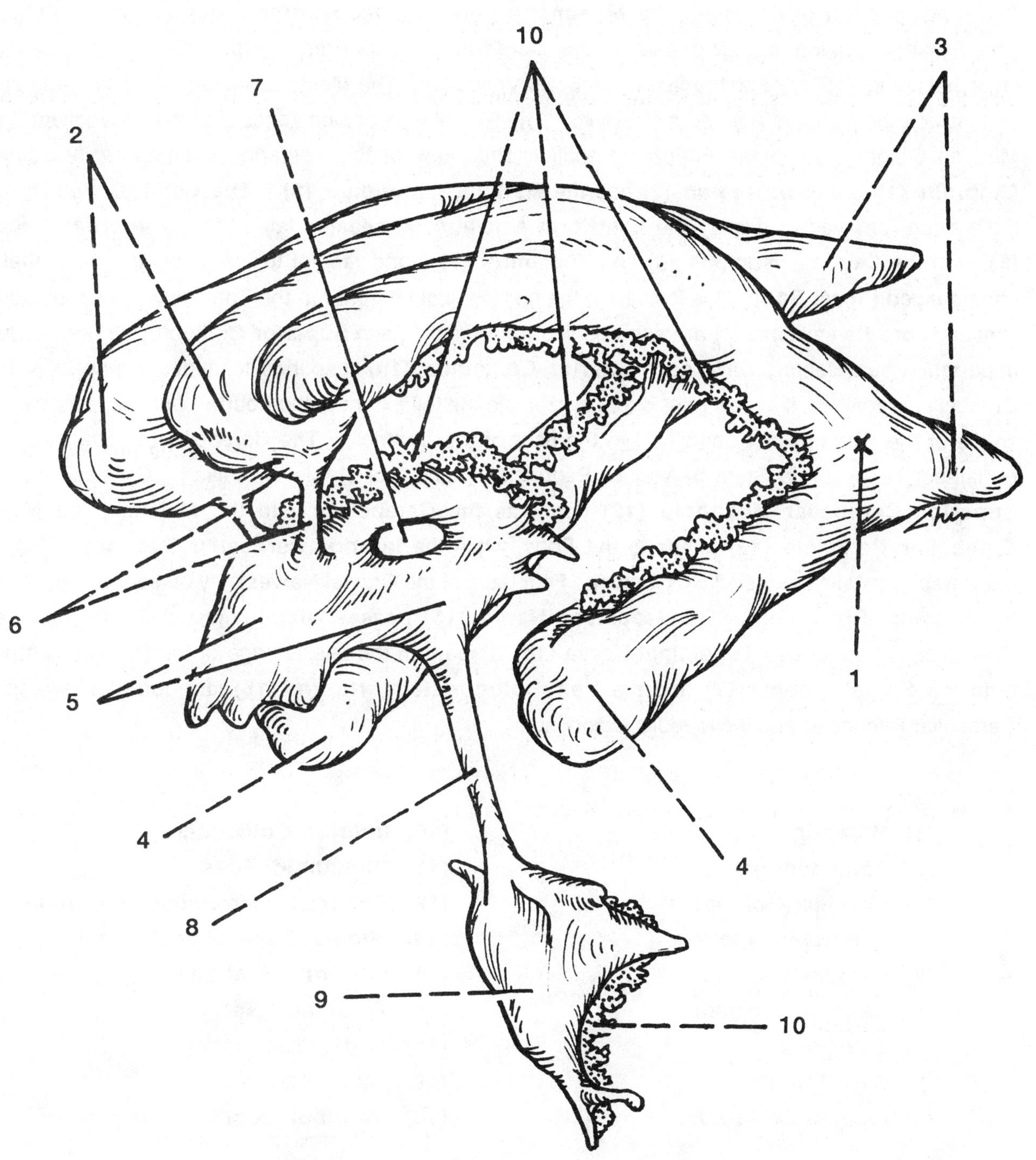
10
3
7
2
6
5
1
4
4
8
9
10
Chin

Posterior Brain Stem

The anterior end of the embryonic Neural Tube expands to form three primary Brain vesicles: the **Prosencephalon** or Forebrain, the **Mesencephalon** and the **Rhombencephalon** or Hindbrain. The Prosencephalon will give rise to the Cerebral Hemispheres, Corpus Callosum, Thalamus, Hypothalamus, Pituitary and the first three Brain Ventricles. The Mesencephalon will form the Midbrain and include the Cerebral Aqueduct of Sylvius. The Rhombencephalon consists of the Cerebellum, Pons, Medulla Oblongata and the Fourth Ventricle. This view of the Posterior Brain Stem includes the **Midbrain (1)**, the **Pons region (2)** and the **Medulla Oblongata (3)**. The Cerebral Hemispheres have been removed to display the **Lentiform Nucleus (4)** situated lateral to the **Internal Capsule (5)**, with the **Caudate Nucleus (6)** and **Thalamus (7)** found medial to the Capsule. Note that the cone-shaped **Pineal Body (8)** is located in the midline, just posterior to the Thalamus. The four Colliculi ("mounds or little eminences") are seen behind the Pineal. Each **Superior Colliculus (9)** serves as an important visual center, while each **Inferior Colliculus (10)** responds to auditory input. With the Cerebellum removed, it is now possible to appreciate the full extent of the Fourth Ventricle, especially the shape of the **Rhomboid Fossa (11)** in the floor of the Ventricle. The Cerebellum is attached to the remainder of the Brain Stem by way of Peduncles ("little foot stems or stalks"). On each side, the **Superior Cerebellar Peduncle (12)** connects the Cerebellum with the Midbrain, the **Middle Cerebellar Peduncle (13)** unites to the Pons, while the **Inferior Cerebellar Peduncle (14)** links the Cerebellum with the Medulla Oblongata. Four pairs of the Cranial Nerves may be seen emerging from the Posterior Brain Stem. The **Trochlear Nerve (15)** passes laterally, just distal to the Inferior Colliculus. The massive **Trigeminal Nerve (16)** is seen on the lateral edge of the Pons region, while both the **Facial Nerve (17)** and the **Vestibulocochlear Nerve (18)** are found between the Cerebellar Peduncles and the Medulla Oblongata.

(1) Midbrain
(2) Pons region
(3) Medulla Oblongata
(4) Lentiform Nucleus
(5) Internal Capsule
(6) Caudate Nucleus
(7) Thalamus
(8) Pineal Body
(9) Superior Colliculus
(10) Inferior Colliculus
(11) Rhomboid Fossa
(12) Superior Cerebellar Peduncle
(13) Middle Cerebellar Peduncle
(14) Inferior Cerebellar Peduncle
(15) Trochlear Nerve
(16) Trigeminal Nerve
(17) Facial Nerve
(18) Vestibulocochlear Nerve

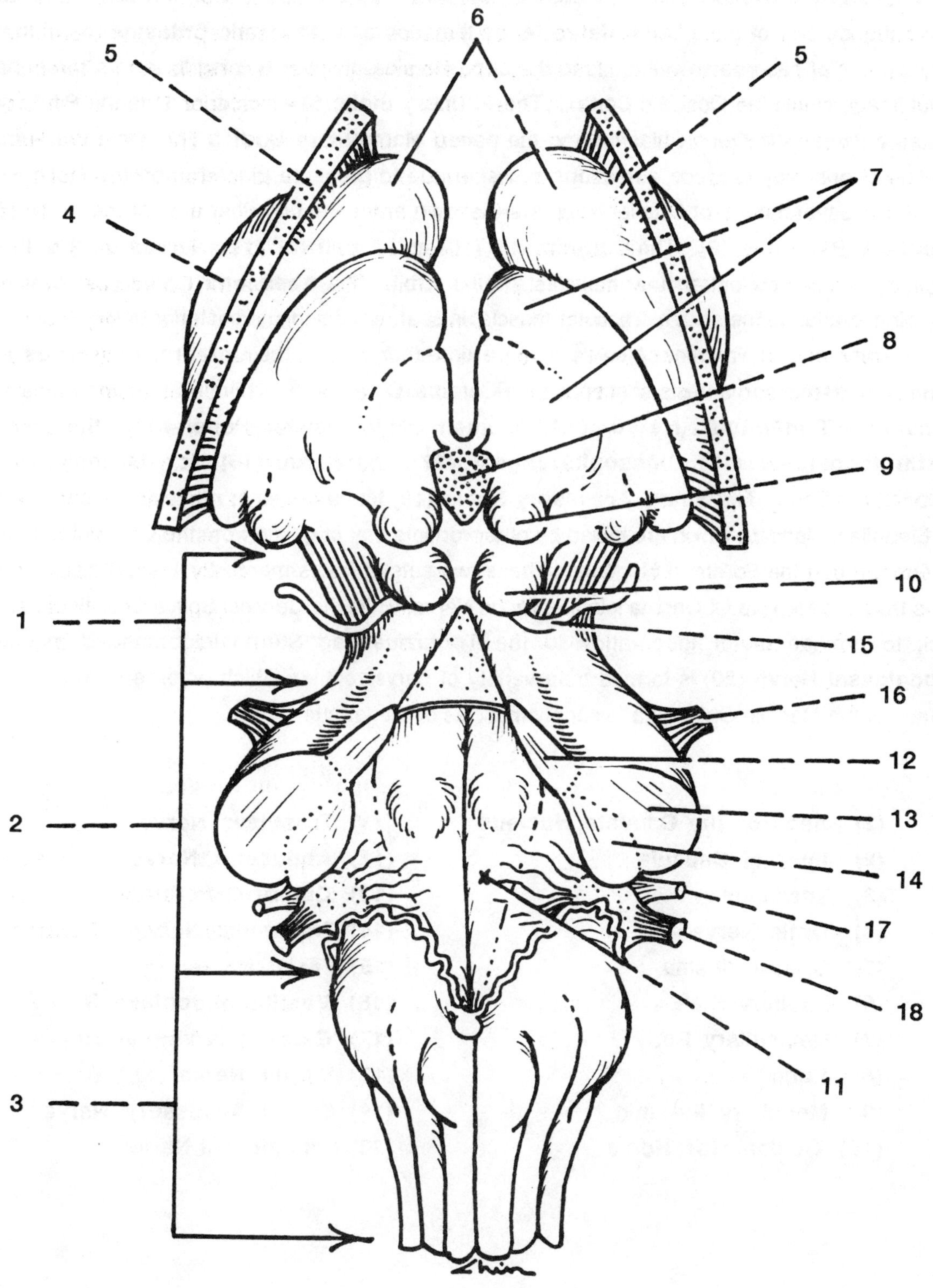

NS-11

Anterior Brain Stem

This anterior (or ventral) view of the Brain Stem is made possible by removal of the Cerebral Hemispheres and the Cerebellum. Beginning at the rostral end, the **Head of the Caudate Nucleus (1)** is seen on either side of the midline, with the **Internal Capsule (2)** situated just lateral to the **Thalamus (3)**. Notice the location of each **Optic Nerve (4)** as they approach the **Optic Chiasma (5)** where the most medial fibers of each nerve will cross to the opposite side, thereby helping to form a three dimensional visual image within the Occipital Cortex. The **Pituitary Stalk (6)** which suspends the Pituitary Gland is located between the Optic Chiasma and the paired **Mammillary Bodies (7)**. Note that tumors of the Pituitary Gland may produce alterations in vision due to pressure placed upon the Optic Nerves and Chiasma. Several pairs of Cranial Nerves are shown emerging on either side of the **Pons (8)** and the **Medullary Pyramids (9)**. The **Oculomotor (10)** and **Trochlear (11) Nerves** pass to the Eye and supply five of the six extraocular muscles in each orbit. The **Abducens nerve (12)**, which provides motor innervation to the sixth extraocular muscle, is seen emerging just posterior to the Pons. Notice that the Oculomotor and Trochlear Nerves emerge on either side of the **Cerebral Peduncles (13)** which represent massive nerve tracts connecting the Cerebral Cortex with the Brain Stem and Spinal Cord. The prominent **Trigeminal Nerve (14)** is seen on the lateral Pons, while the **Facial (15)**, **Vestibulocochlear (16)**, **Glossopharyngeal (17)**, and **Vagus (18) Nerves** are associated with the posterior Pons. The **Spinal Accessory Nerve (19)** has a series of nerve rootlets which arise from the Medulla Oblongata which are joined by other rootlets that originate from the Cervical Spinal Cord and ascend through the Foramen Magnum. These two sets of fibers merge for a short distance before the fibers from the Medulla Oblongata join the Vagus Nerve, while the Cervical Spinal Cord fibers pass into the neck to provide motor innervation to the Trapezius and Sternocleidomastoid muscles. The **Hypoglossal Nerve (20)** is formed by a variety of nerve rootlets which emerge from the ventrolateral surface of the Medulla Oblongata to supply muscles of the tongue.

(1) Head of the Caudate Nucleus
(2) Internal Capsule
(3) Thalamus
(4) Optic Nerve
(5) Optic Chiasma
(6) Pituitary Stalk
(7) Mammillary Body
(8) Pons
(9) Medullary Pyramid
(10) Oculomotor Nerve
(11) Trochlear Nerve
(12) Abducens Nerve
(13) Cerebral Peduncles
(14) Trigeminal Nerve
(15) Facial Nerve
(16) Vestibulocochlear Nerve
(17) Glossopharyngeal Nerve
(18) Vagus Nerve
(19) Spinal Accessory Nerve
(20) Hypoglossal Nerve

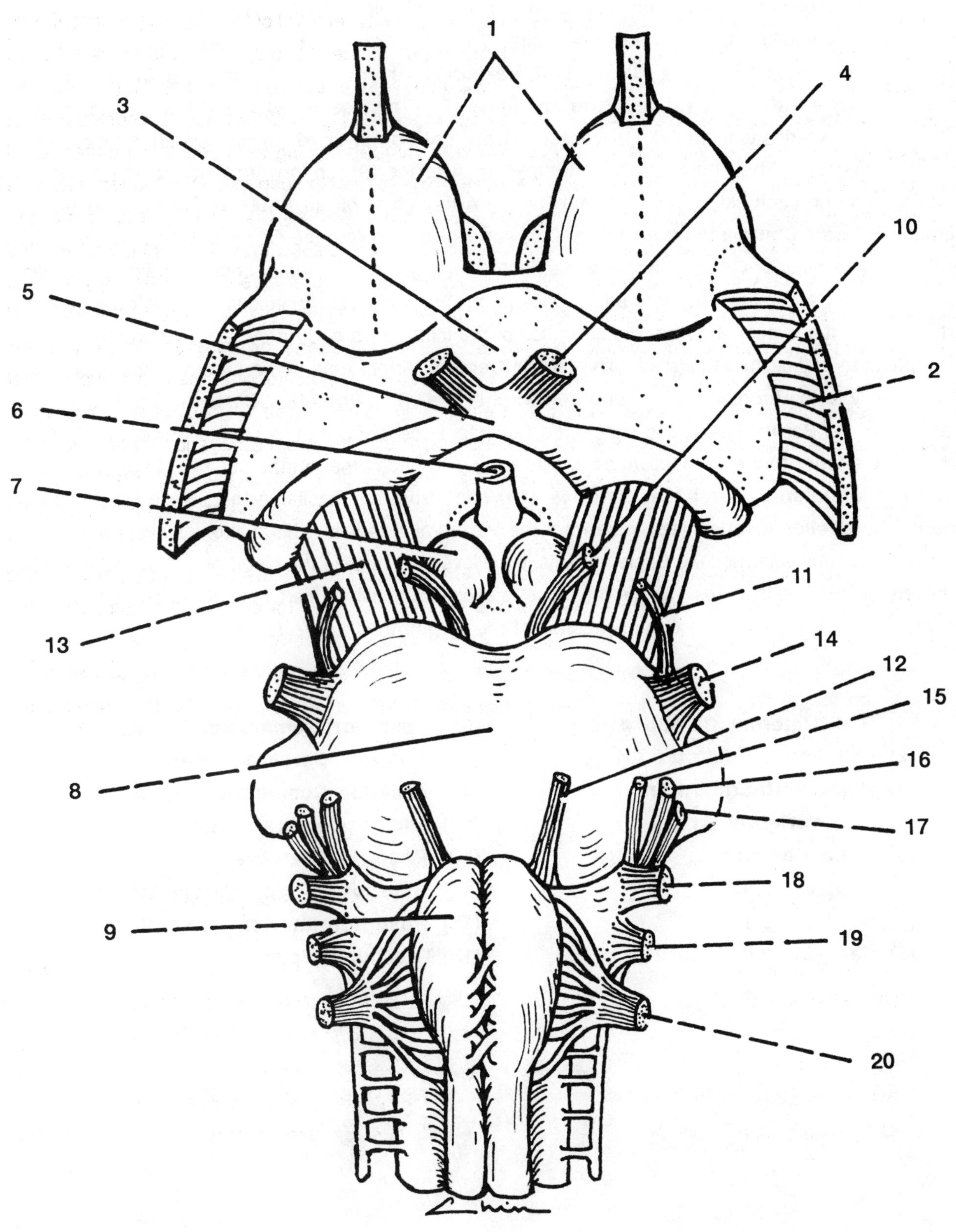

NS-12

Interior of the Skull Relations

The *in situ* relationships of the Arterial blood supply to the Brain is best appreciated with a view of the base of the skull following removal of the brain. The **Right (1)** and **Left Internal Carotid Arteries (2)** anastomose with the **Right (3)** and **Left Vertebral Arteries (4)** forming the **Circle of Willis** around the **Optic Chiasma (5)** and **Pituitary Stalk (6)**. The **Basilar Artery (7)** is formed from the two Vertebral Arteries which have followed the **Cervical Spinal Cord (8)** and ascended through the Foramen Magnum. The **Posterior Communicating Arteries (9)** run forward to meet with the Internal Carotid Artery on each side. Each Internal Carotid Artery gives off an **Anterior Cerebral Artery (10)** with the **Anterior Communicating Artery (11)** serving to complete the Circle of blood flow described by Willis. The Anterior Cerebral Artery is distributed throughout the Frontal Lobe and along the medial surface of the Cerebral Hemisphere as far back as the Parieto-Occipital sulcus. The **Middle Cerebral Artery (12)** is the largest and most direct branch of the Internal Carotid Artery, and is therefore the most susceptible to embolism and stroke. The Middle Cerebral Artery runs in the Lateral Fissure to provide blood supply to the Frontal, Parietal and Temporal Lobes. The **Posterior Cerebral Arteries (13)** are branches of the Basilar Artery, and supply the Occipital Lobe along with the remainder of the Temporal Lobe. The arteries to the Cerebellum all anastomose with each other on the surface of the Cerebellum. The **Anterior Inferior Cerebellar Arteries (14)** arise from the Basilar Artery, while the **Posterior Inferior Cerebellar Arteries (15)** arise from the Vertebral Arteries. This elaborate system of arterial anastomoses, that is termed the Circle of Willis, is believed to equalize the blood pressure and blood flow between the two hemispheres. Notice the close relationship between the Circle of Willis and the Cranial Nerves pictured, including the three branches of the **Trigeminal Nerve (16).**

(1) Right Internal Carotid Artery
(2) Left Internal Carotid Artery
(3) Right Vertebral Artery
(4) Left Vertebral Artery
(5) Optic Chiasma
(6) Pituitary Stalk
(7) Basilar Artery
(8) Cervical Spinal Cord
(9) Posterior Communicating Artery
(10) Anterior Cerebral Artery
(11) Anterior Communicating Artery
(12) Middle Cerebral Artery
(13) Posterior Cerebral Artery
(14) Anterior Inferior Cerebellar Ar.
(15) Posterior Inferior Cerebellar Ar.
(16) Trigeminal Nerve

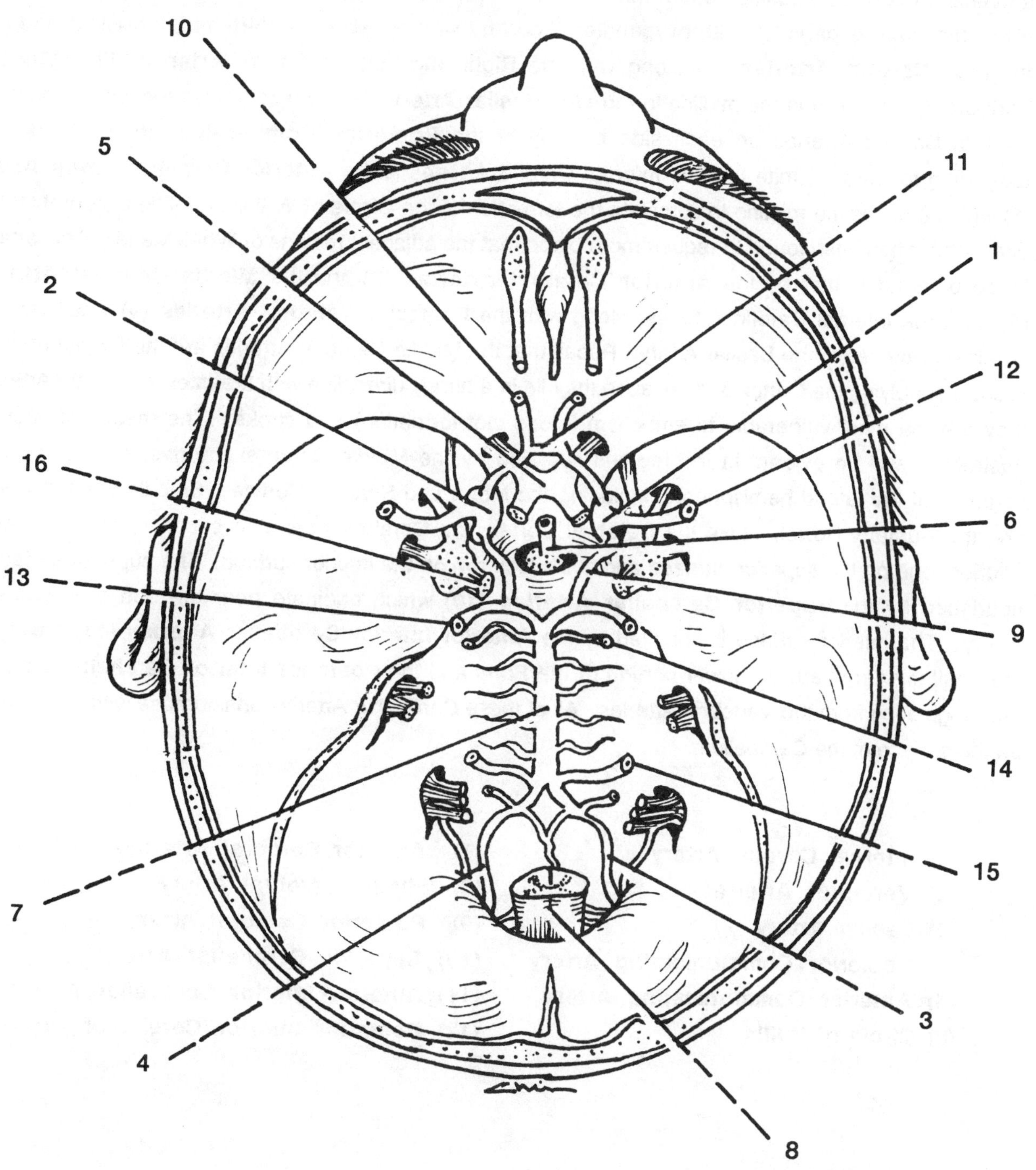
10
5
11
1
2
12
16
6
13
9
14
15
7
3
4
8

Cerebral Arterial Circle of Willis

(Inferior View)

This is a view of the Circle of Willis seen from below with all of the adjacent brain tissue removed in order to appreciate the rich branching pattern that is evident in this arterial circulation. As noted in the previous plate, the brain is provided with oxygenated blood by way of two paired arteries, the Right and Left **Internal Carotid Arteries (1)** along with the Right and Left **Vertebral Arteries (2)**. The two Vertebral Arteries join in the midline to form the **Basilar Artery (3)** which continues forward to meet the Internal Carotid Arteries on each side by way of the **Posterior Communicating Arteries (4)**. Extending forward to unite the two Internal Carotid Arteries is the **Anterior Communicating Artery (5)** which crosses the midline to complete the formation of the **Circle of Willis (6)**. The Cerebral Cortex Gray Matter has been found to require more blood that the adjacent regions of White Matter. This arterial blood demand is met by the **Anterior Cerebral Arteries (7)** and the **Middle Cerebral Arteries (8)** from the Internal Carotid Arteries, along with the **Posterior Cerebral Arteries (9)** which are the terminal branches of the Basilar Artery. Recall that the Middle Cerebral Arteries are the largest of these vessels supplying the Cortex and because they lie in a nearly direct line with the Internal Carotid Arteries, they are the most vulnerable to embolism (blood clot formation) and stroke. The results of such an obstruction will be evident in the regions supplied by the Middle Cerebral Arteries, namely a major segment of the lateral hemispheres, including the Motor and Sensory Cortices, Broca's Speech Center and the Auditory Cortex. Like the Cerebral Cortex, the Cerebellum is also supplied by three paired arteries, one on the superior surface and two which run on the inferior surface. The superior surface is nourished by the **Superior Cerebellar Arteries (10)** which originate near the end of the Basilar Artery. The inferior surface is supplied by the **Anterior Inferior Cerebellar Arteries (11)** arising off the Basilar Artery near the lower portion of the Pons and the **Posterior Inferior Cerebellar Arteries (12)** originating from the Vertebral Arteries. All of these Cerebellar Arteries anastomose with one another on the surface of the Cerebellum.

(1) Internal Carotid Artery
(2) Vertebral Arteries
(3) Basilar Artery
(4) Posterior Communicating Artery
(5) Anterior Communicating Artery
(6) Circle of Willis

(7) Anterior Cerebral Arteries
(8) Middle Cerebral Artery
(9) Posterior Cerebral Artery
(10) Superior Cerebellar Artery
(11) Anterior Inferior Cerebellar Artery
(12) Posterior Inferior Cerebellar Artery

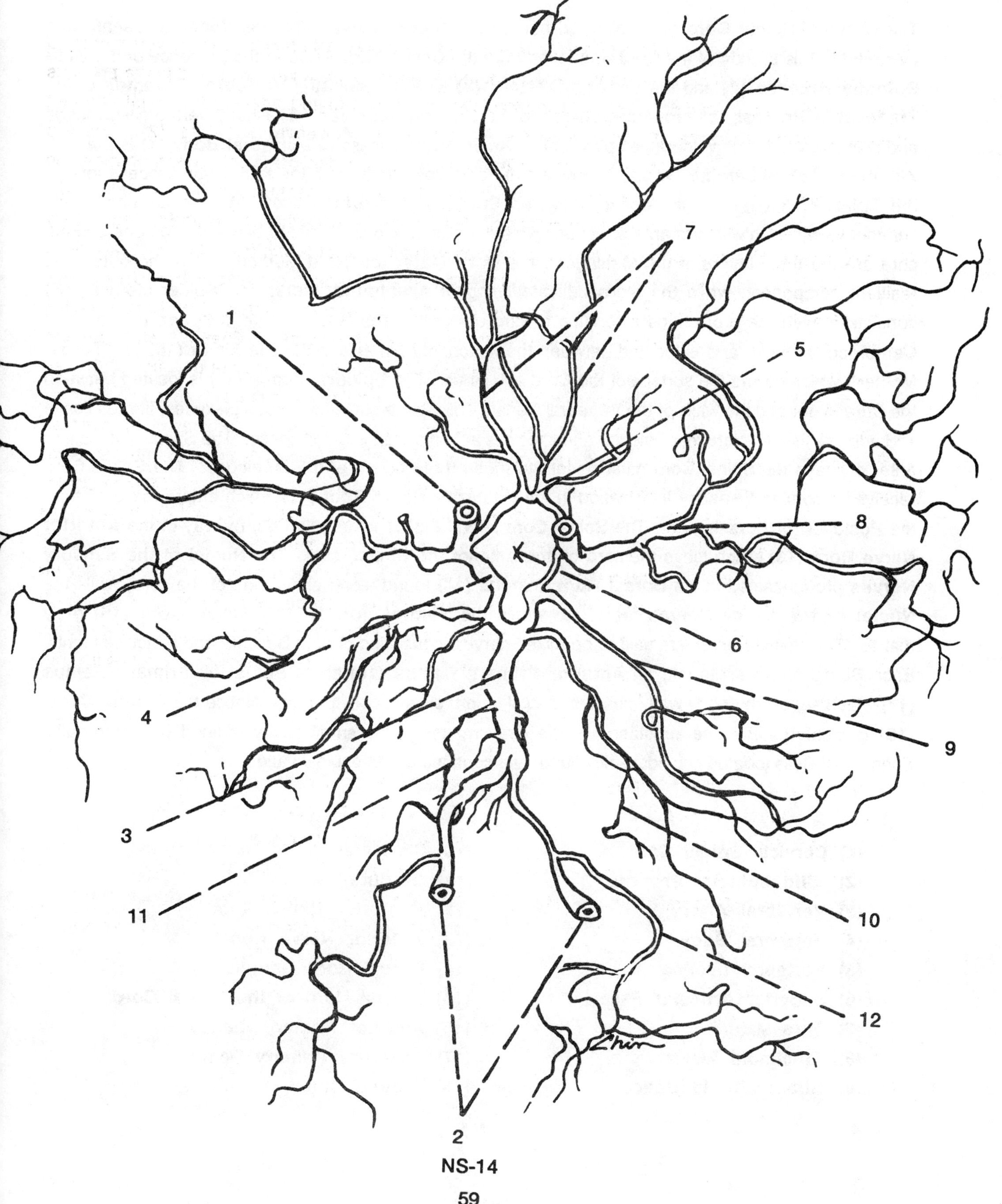

NS-14

Cross Section of the Spinal Cord

The **Cervical Spinal Cord (1)** is pictured here as seen from above within the Vertebral Canal. It is evident that this section of the Cord was taken from the Cervical Level due to the presence of the **Bifid Spinous Process (2)** and the **Vertebral Artery (3)** located within the **Vertebral Foramen** of the **Transverse Process**, all properties unique to the Cervical Vertebrae. The bony vertebra surrounds and protects the rather vulnerable spinal cord. Anteriorly, the massive **Vertebral Body (4)** is present, with the **Vertebral Lamina (5)** on each side extending posteriorly from the transverse process toward the Spinous Process. The cartilage-covered **Superior Articular Facet (6)** is also seen in this superior view, which will form an articulation with the vertebra above. Situated between the bone and the cord are the three layers of the Meninges: the **Dura Mater** (or "Hard Mother") **(7)** is the outermost resilient component, while the more delicate lining of **Arachnoid Mater** (or "Spider Mother") **(8)** consists of web-like projections extending toward the cord. The **Subarachnoid Space (9)** contains Cerebrospinal Fluid, and is located between the Arachnoid Mater and the **Pia Mater (10)** (or "Tender Mother") which covers the surface of the Cord and Brain. The **Epidural Space (11)** is located between the Dura Mater and the walls of the Vertebral Canal. This space contains venous plexuses, fibrous tissue and adipose tissue. Note that Epidural anesthesia is administered into this space. The Spinal Nerves that enter or leave the Spinal Cord must penetrate these three layers of the Meninges. The Spinal Cord receives incoming **Sensory** information in the **Posterior Nerve Root (12)** which enters the cord from the Posterior (or Dorsal) side. The Spinal Cord sends **Motor** information out by way of the **Anterior Nerve Root (13)** which leave the cord on the Anterior (or Ventral) side. The **Nuclei of the Sensory Nerves** are contained in the **Dorsal Root Ganglia (14)** found along either side of the cord, while the **Nuclei of the Motor Nerves** are found within the **Ventral Horn of the Spinal Cord (15)** gray matter. The Spinal Nerves are said to be mixed nerves, consisting of both Sensory and Motor elements. Each Spinal Nerve splits into an **Anterior Primary Ramus (16)** and a **Posterior Primary Ramus (17)** which supply the body wall, muscles, and skin in their respective areas. Notice the **Central Canal (18)** is located within the substance of the Gray matter. This small cavity is lined with Neuroglial Ependymal Cells (ciliated cuboidal epithelium) and contains Cerebrospinal Fluid.

(1) Cervical Spinal Cord
(2) Bifid Spinous Process
(3) Vertebral Artery
(4) Vertebral Body
(5) Vertebral Lamina
(6) Superior Articular Facet
(7) Dura Mater
(8) Arachnoid Mater
(9) Subarachnoid Space
(10) Pia Mater
(11) Epidural Space
(12) Posterior Nerve Root
(13) Anterior Nerve Root
(14) Dorsal Root Ganglion
(15) Ventral Horn of the Spinal Cord
(16) Anterior Primary Ramus
(17) Posterior Primary Ramus
(18) Central Canal

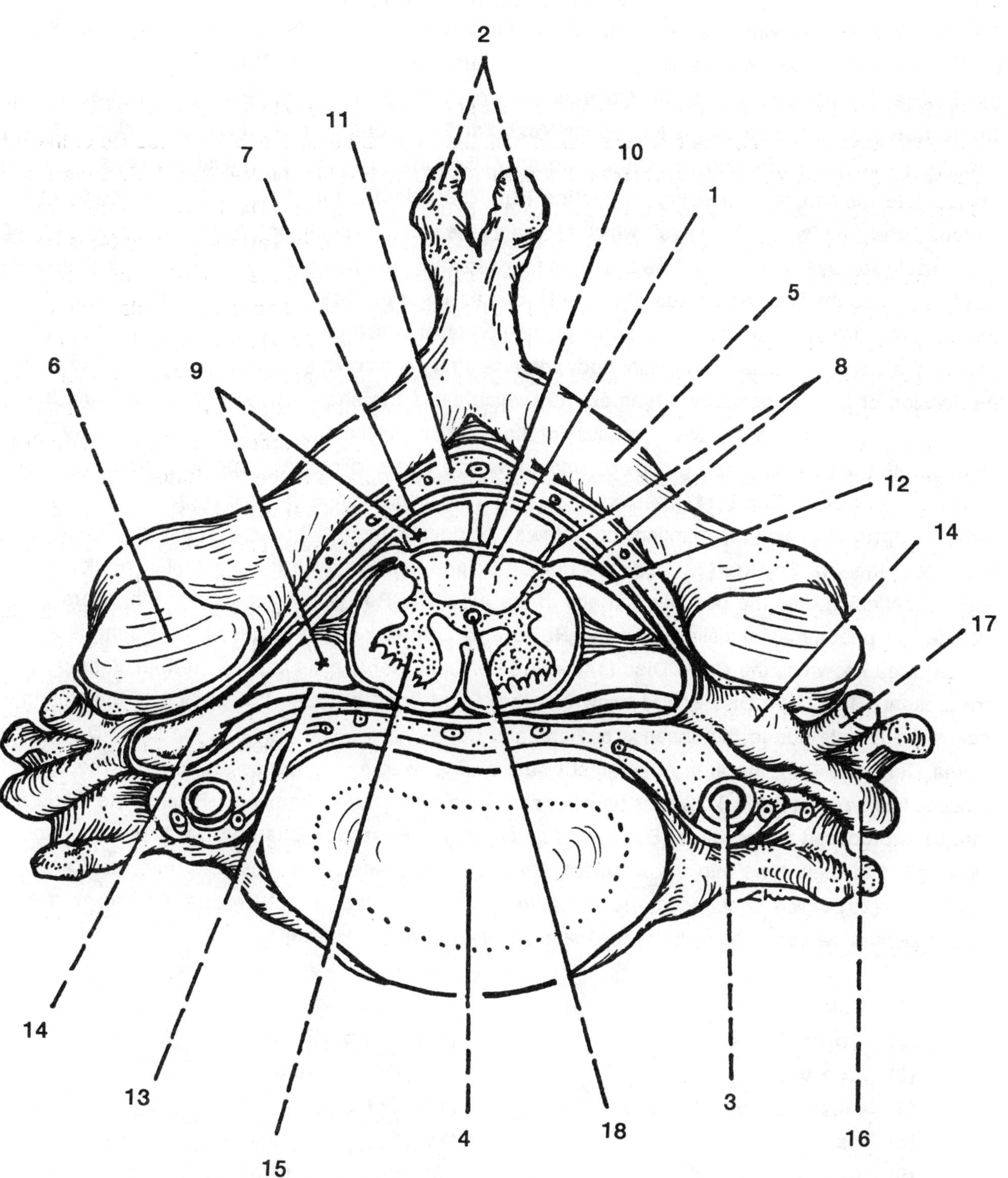

NS-15

The Eye

The Eyeball is located within the bony orbit of the skull and is surrounded by protective fat. The Wall of the Eyeball is comprised of three coats or tunics: The outermost coat is the **Fibrous Tunic** consisting of the **Sclera (1)** posteriorly and the **Cornea (2)** anteriorly, which is covered by the transparent **Conjunctiva**. The middle coat is termed the **Vascular Tunic** due to the abundance of blood vessels contained within the **Choroid (3), Ciliary Body (4)** and **Iris (5)**. The innermost **Nervous Tunic** is the delicate membrane comprising the **Retina (6)**. The Sclera is the posterior opaque region of the Fibrous Tunic (the "white of the eye") which serves as the insertion point for the six extraocular muscles responsible for eye movement. Two of these extraocular muscles are shown here, the **Superior Rectus (7)** and the **Inferior Rectus (8)**. The fibrous tissue of the Sclera projects anteriorly to form the transparent Cornea. The Choroid is a thin pigmented layer which provides a rich capillary network to nourish the adjacent Retina. The Ciliary Body contains a ring of smooth muscle fibers which act to control the tension of the **Suspensory Ligament (9)** which alters the shape of the **Lens (10)**. The **Iris** is the colored portion of the eye which consists of the anterior projection of the Choroid and Ciliary Body. Note that the Iris contains two groups of smooth muscle fibers (dilator and sphincter muscles) which control the size of the **Pupil (11)** opening. The Retina represents an embryonic outgrowth of the Brain, with the **Optic Nerves (12)** marking the pathway of development. The light-sensitive Retina ends abruptly at the **Ora Serrata (13)** near the Corneo-Scleral junction, with the pigmented portion of the Retina continuing onto the posterior surface of the Iris. The Retina is composed of ten layers which include the photoreceptor cells called the **Rods and Cones**. At the center of the Retina is a pale circular area known as the **Optic Disc (14)** where the Optic Nerve exits and the **Retinal Vessels (15)** are present. The Optic Disc is also known as the "Blind Spot" because this portion of the Retina is insensitive to light due to the absence of Rods and Cones here. Just lateral to the Optic Disc is the **Fovea Centralis (16)**, the area of greatest visual acuity, where only Cones are found. The Eyeball contains two types of fluids, known as humors, contained within three chambers: Plasma-like **Aqueous Humor** is produced by the Ciliary Body and first enters the **Posterior Chamber (17)**, passes through the Pupil, and then flows into the **Anterior Chamber (18)** where it is absorbed into the **Canal of Schlemm (19)** which is connected to the anterior scleral veins. The **Vitreous Chamber (20)** is located behind the Lens and contains the clear, jelly-like **Vitreous Humor**.

(1) Sclera
(2) Cornea
(3) Choroid
(4) Ciliary Body
(5) Iris
(6) Retina
(7) Superior Rectus muscle
(8) Inferior Rectus muscle
(9) Suspensory Ligament
(10) Lens
(11) Pupil
(12) Optic Nerve
(13) Ora Serrata
(14) Optic Disc
(15) Retinal Vessels
(16) Fovea Centralis
(17) Posterior Chamber
(18) Anterior Chamber
(19) Canal of Schlemm
(20) Vitreous Chamber

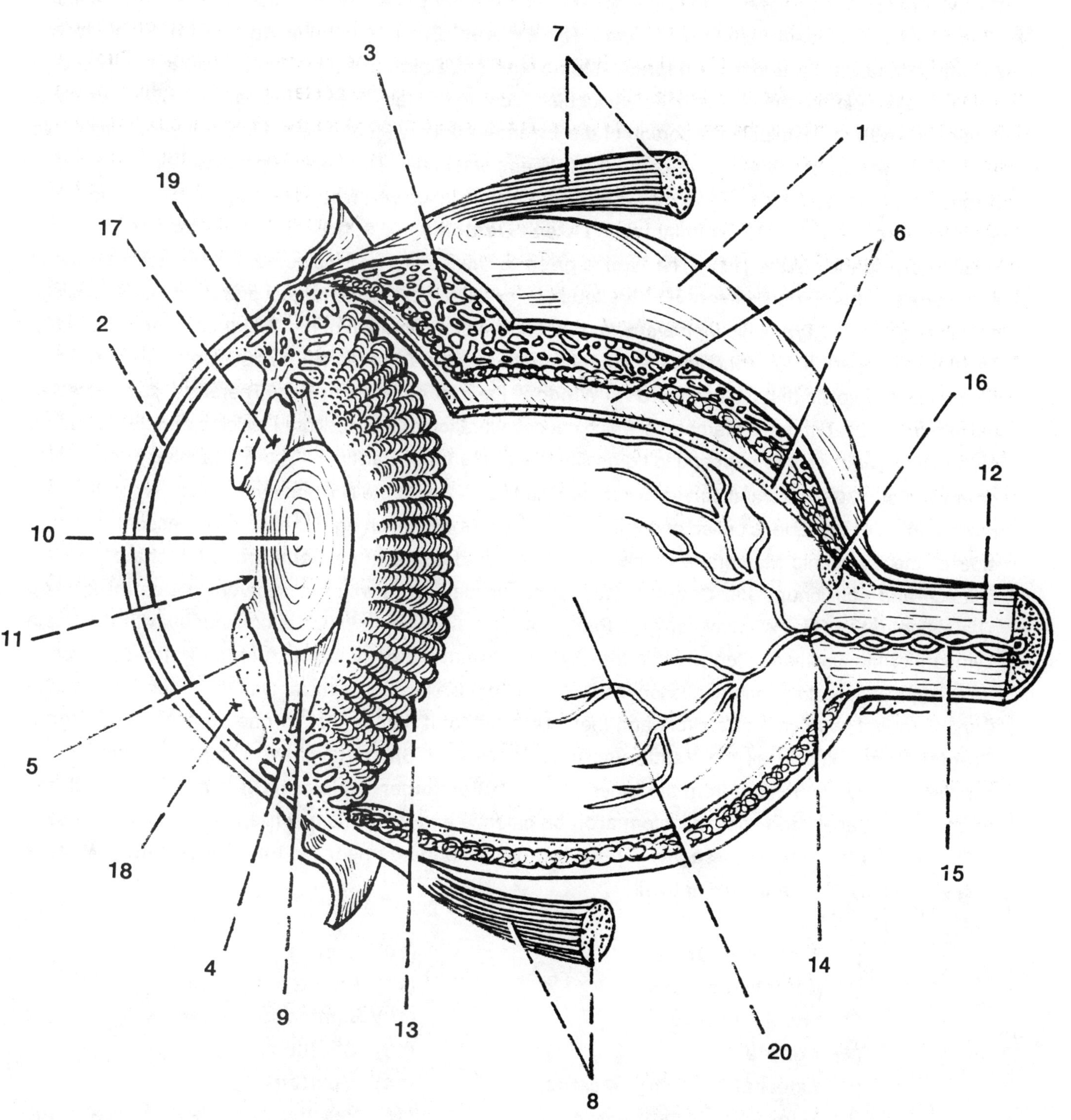
7
3
1
19
17
6
2
16
12
10
11
5
18
15
14
4
9
13
20
8

The Ear

The Ear is housed within the protective **Petrous** ("stony") portion of the **Temporal Bone (1)**, with the **Mastoid Process (2)** and **Styloid Process (3)** of the Temporal Bone extending below the Ear Cavity. The Ear is composed of the External, Middle and Inner Ear elements which are concerned with the special senses of hearing and equilibrium. The **External Ear** includes the **Auricle (4)**, the **External Auditory Meatus (5)**, and the **Tympanic Membrane** (or "Eardrum") **(6)**. The Auricle is composed of resilient yellow elastic cartilage, with the ear lobule filled with soft fatty tissue. The cartilage of the Auricle is continuous with the cartilaginous portion of the External Auditory Meatus found in the lateral third, while the inner two-thirds of the Meatus is composed of Temporal bone. The Meatus includes hairs, sebaceous glands, and modified sweat glands known as Ceruminous glands which secrete a yellow-brown wax. The **Middle Ear Cavity (7)** contains the **Three Ossicles (8,9,10)** and the **Eustachian** (or Pharyngotympanic) **Tube (11).** The names given to the Auditory Ossicles reflect the shape of these smallest elements of the skeleton. The **Malleus (8)** (or "Hammer") is the largest and most lateral Ossicle, articulating with the Tympanic Membrane and transmitting the vibration of the "Eardrum" to the **Incus (9)**. The "Anvil-shaped" Incus conducts the sound vibration to the **Stapes (10)** (or "Stirrup") which includes a prominent footplate that fits into the **Oval Window**, the inlet of the Inner Ear. Therefore, sound creates movement of the Tympanic Membrane which causes vibration of the Auditory Ossicles and displacement of the Oval Window, with corresponding ossilations of the Inner Ear fluid. The Eustachian Tube slopes downward at a 30° angle and serves to connect the Middle Ear to the Nasopharynx. With swallowing, the lower portion of the tube is pulled and the tube opens (arrow), which equalizes the air pressure on either side of the Tympanic Membrane. The Inner Ear includes the three **Semicircular Canals (12)** for balance and equilibrium, the **Cochlea (13)** for sound detection, with the intervening **Vestibule (14)**. The inside of the Oval Window contains Perilymph of the Vestibule which is continuous with the Scala Tympani of the Cochlea. Perilymph is also found between the bony labyrinth of the Petrous Temporal Bone and the delicate membranous labyrinth which comprise the Semicircular Canals. Each Semicircular Canal is about two-thirds of a circle and they are situated at right angles to one another. The **Eighth Cranial Nerve**, known as the **Vestibulocochlear Nerve (15)**, contains two distinct divisions which convey sensory information from the Inner Ear. The **Vestibular Division (16)** is associated with the Semicircular Canals and contains information on balance and equilibrium, while the **Cochlear Division (17)** transmits the sound messages. These two divisions merge near the **Internal Auditory Meatus (18)** where they exit the Inner Ear Cavity.

(1) Temporal Bone
(2) Mastoid Process
(3) Styloid Process
(4) Auricle
(5) External Auditory Meatus
(6) Tympanic Membrane
(7) Middle Ear Cavity
(8) Malleus
(9) Incus
(10) Stapes
(11) Eustachian Tube
(12) Semicircular Canals
(13) Cochlea
(14) Vestibule
(15) Vestibulocochlear Nerve (VIII)
(16) Vestibular Division of VIII
(17) Cochlear Division of VIII
(18) Internal Auditory Meatus

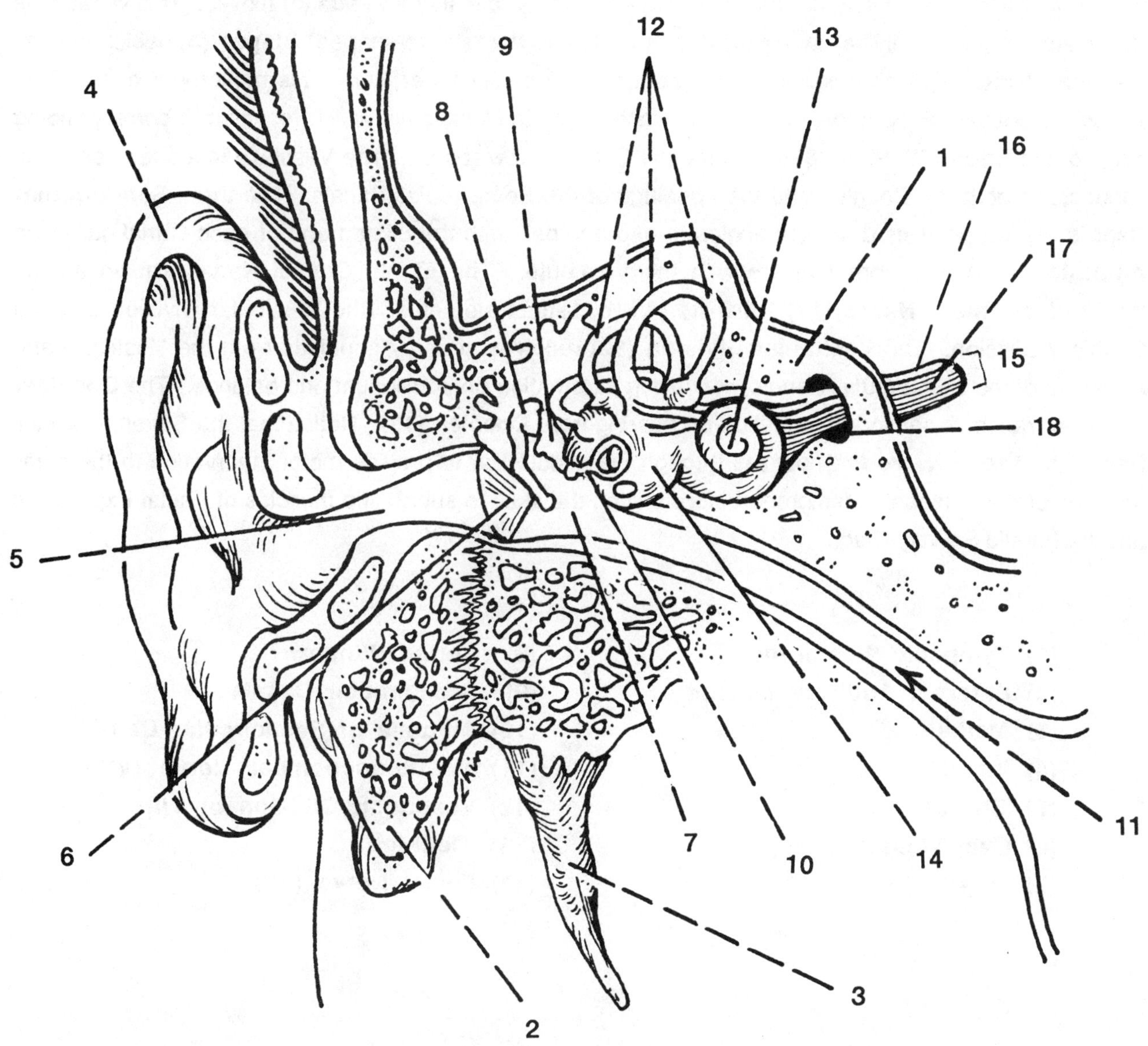
12
9
13
4
8
1
16
17
15
18
5
6
7
10
14
11
2
3

Middle and Inner Ear

The various elements of the Middle and Inner Ear are contained within the **Petrous portion of the Temporal Bone**. The **Tympanic Membrane (1)** is located at the lateral border of the Middle Ear Cavity and is attached to the bony edge of the **External Auditory Meatus (2)**. Sound causes vibration of the Tympanic Membrane which results in movement of the Auditory Ossicles contained within the Middle Ear Cavity. First, the "hammer-shaped" Ear Ossicle, the **Malleus (3)** moves. This vibration is transmitted to the "Anvil-like" **Incus (4)** and ultimately to the "stirrup-shaped" **Stapes (5)**, which causes the **Oval Window (6)** to bulge into the **Vestibule (7)** of the Inner Ear. This movement of the Oval Window produces displacement of the Perilymph within the Vestibule, which results in a corresponding bulge of the **Round Window (8)** each time the Oval Window moves. The Vestibule is a chamber which connects to both the Cochlea and the openings of the Semicircular Canals. The three **Semicircular Canals (9)** are positioned at right angles to one another, with the dilated end of each Canal called an **Ampulla (10)** which communicate with the Vestibule. The Eighth Cranial Nerve, known as the **Vestibulocochlear Nerve (11)**, consists of two distinct elements: the Vestibular Division and the Cochlear Division. The **Vestibular Division (12)** receives sensory impulses from the Vestibule and Ampullae of the Semicircular Canals, conveying information on equilibrium and balance. The **Cochlear Division** transmits sensory information on hearing from **Cochlea (13)**. Notice that the Seventh Cranial Nerve, the **Facial Nerve (14)**, passes through the Middle Ear to provide motor innervation to the small muscles of the Tympanic Membrane before exiting the skull to supply the muscles of Facial Expression and the Parotid Salivary Gland.

(1) Tympanic Membrane
(2) External Auditory Meatus
(3) Malleus
(4) Incus
(5) Stapes
(6) Oval Window
(7) Vestibule
(8) Round Window
(9) Semicircular Canals
(10) Ampulla of Semicircular Canal
(11) Vestibulocochlear Nerve (VIII)
(12) Vestibular Division of VIII
(13) Cochlea
(14) Facial Nerve (VII)

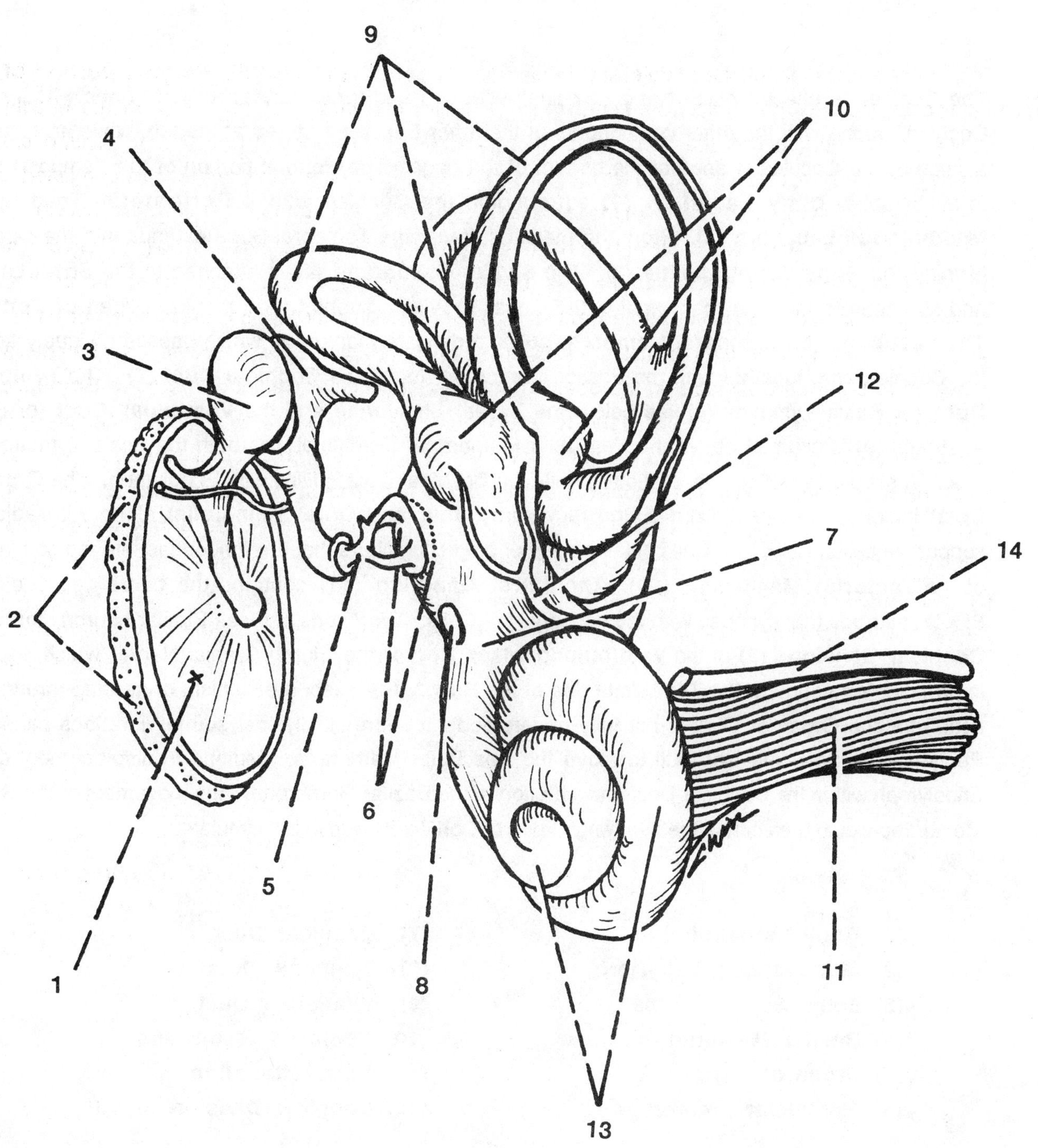
9
10
4
3
12
7
14
2
6
5
1
8
11
13

Cochlea of the Inner Ear

The Cochlea is shaped much like a sea shell with two and three-quarter turns of bone. The entire Cochlea, along with the other components of the Inner Ear, are pictured at the top, while an enlarged section of the Cochlea is seen at the bottom of the page. The Petrous portion of the Temporal Bone forms a spiral **Bony Labyrinth (1)** surrounding the Cochlea with a **Periosteum** lined by the **Membranous Labyrinth (2)**. From the medial surface, the Temporal Bone extends into the Cochlea forming the **Bony Spiral Lamina (3)**. The **Basilar Membrane (4)** is attached to the Spiral Lamina and stretches to the outer bony wall of the Canal, providing support for the spiral **Organ of Corti (5)**. The **Vestibular** (or Reissner's) **Membrane (6)** is a delicate membrane which passes obliquely across the Cochlea and, together with the Basilar Membrane, forms the **Cochlear Duct (7)**. The **Tympanic Duct** (or Scala Tympani) **(8)** lies below the Basilar Membrane and the **Vestibular Duct** (or Scala Vestibuli) **(9)** is situated above the Vestibular Membrane. Perilymph fills both the Scala Tympani and Scala Vestibuli, as well as the Vestibule, while the Cochlear Duct is filled with Endolymph. The **Organ of Corti** includes the specialized sound receptors (**Inner and Outer Hair Cells**) along with adjacent supportive cells. Each Hair Cell contains a tuft of projecting fibers that are embedded into the soft lamina of the **Tectorial Membrane (10)**. The **Spiral Ganglion (11)** contains the nerve cell bodies of Bipolar neurons that synapse with the sensory Hair Cells. The Axons of these Bipolar neurons forms the **Cochlear Division (12)** of the **Vestibulocochlear Nerve** (the Eighth Cranial Nerve) which conveys information on sound to the Temporal Lobe of the Brain. The exact mechanism of hearing through the action for the Organ of Corti is not fully understood. It seems likely that sound vibrations cause the Perilymph in the Scala Vestibuli to move the Vestibular Membrane, resulting in displacement of the Endolymph within the Cochlear Duct and vibration of the Basilar Membrane. Any movement of the Basilar Membrane would then cause the overlying Organ of Corti Hair Cells to be stimulated.

(1) Bony Labyrinth
(2) Membranous Labyrinth
(3) Bony Spiral Lamina
(4) Basilar Membrane
(5) Organ of Corti
(6) Vestibular Membrane
(7) Cochlear Duct
(8) Tympanic Duct
(9) Vestibular Duct
(10) Tectorial Membrane
(11) Spiral Ganglion
(12) Cochlear Division of VIII

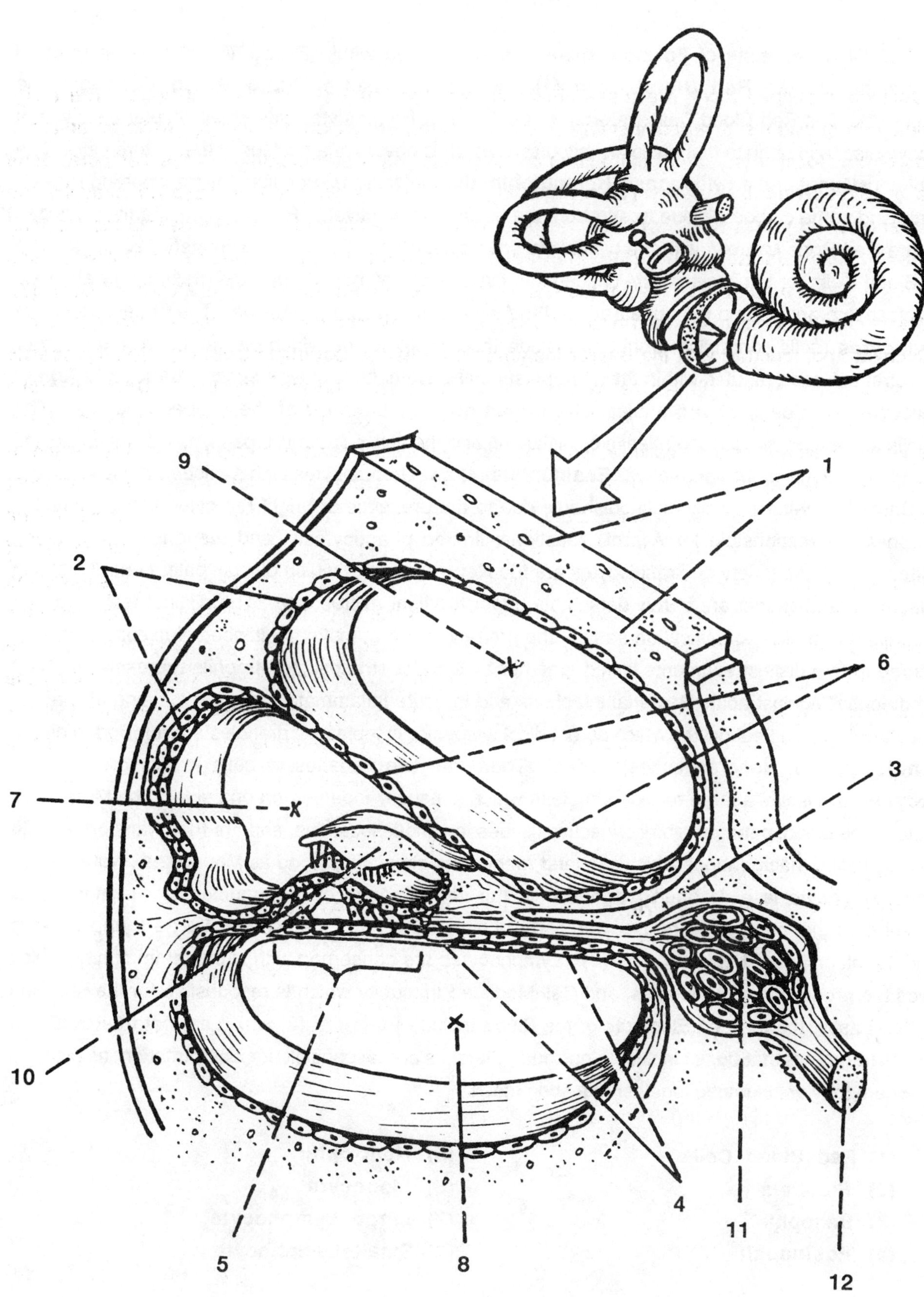
9
1
2
6
3
7
10
4
11
5
8
12

Blood Cells

Circulating Blood consists of **Formed Elements** suspended within **Blood Plasma**. The Formed Elements include the **Red Blood Cells (1)**, the various types of **White Blood Cells** and the **Platelets (2)**. The Red Blood Cells (referred to as RBCs or Erythrocytes) are small circular cells with a biconcave disc appearance that encloses cytoplasm which is devoid of a nucleus in the mature cells. The primary constituent of the cytoplasm is **Hemoglobin**, the molecule responsible for transporting oxygen primarily and some carbon dioxide to and from the tissues, respectively. RBCs are constantly produced within the Red Bone Marrow, with the typical life span of each cell being approximately 120 days. The White Blood Cells (or Leukocytes) are divided into five categories based upon the presence or absence of cytoplasmic granules and the appearance of the stained cytoplasmic granules. The three varieties of **Granulocytes** (cells containing granules) include those Leukocytes which display either blue, red or pale neutral colored granules within their cytoplasm using Romanovsky-type stains. **Basophils (3)** are Granulocytes with deep blue granules which often obscure the view of the bilobed nucleus. The Basophils are responsible for the release of Histamine and the anticoagulant Heparin, as well as displaying limited Phagocytosis of foreign matter. **Eosinophils (4)** are Granulocytes with a bilobed nucleus similar to the Basophils, yet are easily distinguishable due to the presence of bright red cytoplasmic granules. Eosinophils are responsible for Allergic reactions, limited phagocytosis, and responses to invading Parasites. The third variety of Granulocytes are the **Neutrophils (5)** which display pale, neutral staining cytoplasmic granules that are barely detectable under the light microscope. The distinct feature of the Neutrophils that makes identification easy is the presence of a complex multilobed Nucleus containing from three to five distinct segments linked together by slender strands. Neutrophils represent the "first line of defense" against acute Bacterial infections and in acute inflammatory reactions, being efficient in phagocytosis. Those Leukocytes which do not display specific cytoplasmic granules are referred to as the **Agranulocyte** (or Nongranulocyte) series. There are two varieties of cells in this group. The **Monocytes (6)** display a large rounded nucleus which is usually indented on one side. The Monocytes circulate in the blood, penetrate the connective tissues in various locations, and are then transformed into phagocytic tissue Macrophages which can later reenter the circulating blood as Monocytes. Note that an elevated Monocyte level is often diagnostic in the viral disease of Mononucleosis. Lymphocytes represent the other group of Agranulocytes and are pictured here as the younger **Large Lymphocytes (7)** and the older **Small Lymphocytes (8)**. Lymphocytes are concerned with Humoral Immunity, which involves the production of Antibodies, and Cell-Mediated Immunity which is responsible for the rejection of grafts. Lastly, the Formed Elements of the Blood include **Platelets (2)** which are not true cells, but rather cell fragments that do not include a nucleus. Platelets are responsible for clot formation, as pictured here, where they are clumped together in a fibrin mesh.

(1) Red Blood Cells
(2) Platelets
(3) Basophil
(4) Eosinophil
(5) Neutrophil
(6) Monocyte
(7) Large Lymphocyte
(8) Small Lymphocyte

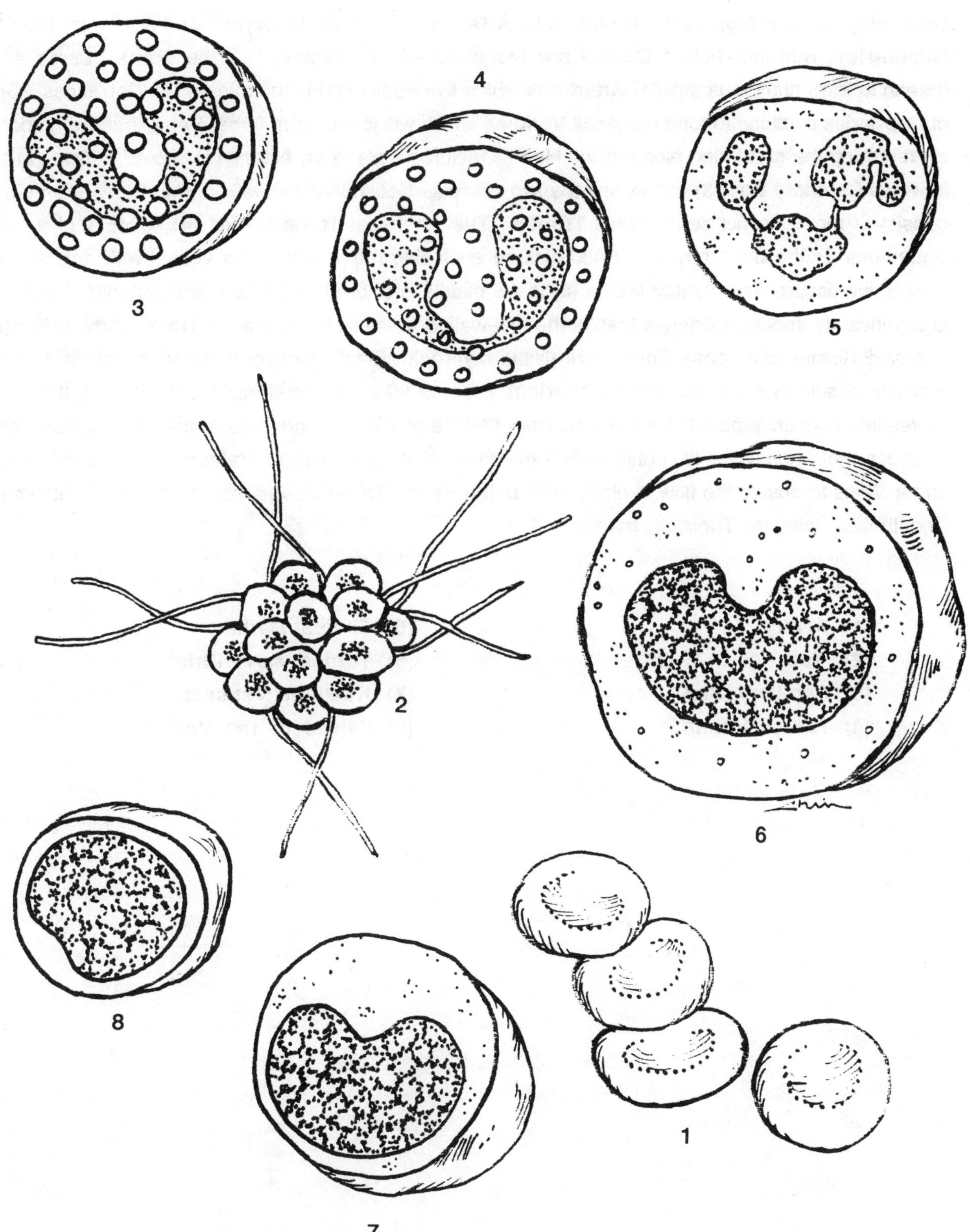

CS-1

Arteries, Capillaries and Veins

Circulating Blood is brought to the tissues by **Arteries** which provide Oxygen and Nutrients though the **Capillaries**, with the Carbon Dioxide and Metabolic Wastes removed by the **Veins**. Large Arteries branch to form numerous smaller **Arterioles** which eventually end in the thin-walled Capillaries. Groups of Capillaries combine to form the small **Venules** which will join to form larger Veins that are responsible for returning the circulating blood to the Heart. Pictured here in an **Artery (1)** above, a **Vein (2)** below it, with a **Capillary Bed (3)** connecting the two vessels. Notice that the walls of the Artery and Vein each consist of three distinct coats called **Tunics**. The innermost **Tunic Intima (4)** consists of a delicate Squamous Epithelium (known as Endothelium) enclosing the lumen of the vessel with adjacent loose connective tissue. The **Tunica Media (5)** is the middle coat of Elastic Fibers and Smooth Muscle which is significantly thicker in Arteries than in the thin-walled Veins. The outermost **Tunica Adventitia (6)** (or Tunica Externa) is a Loose Connective tissue layer with Elastic Collagen fibers and some strands of Smooth Muscle that is penetrated by **Nutrient Vessels (7)** (also known as Vasa Vasorum, the "vessels of vessels") which supply the deeper Smooth Muscle of the Tunica Media in the larger blood vessels. Because the Veins typically contain only a thin layer of Smooth Muscle, **Valves (8)** are found within the larger Veins to ensure the flow of blood back to the Heart. These Valves represent specializations of the Endothelium lining the Tunica Intima.

(1) Artery
(2) Vein
(3) Capillary Bed
(4) Tunica Intima
(5) Tunica Media
(6) Tunica Adventitia
(7) Nutrient Vessels
(8) Valves of the Vein

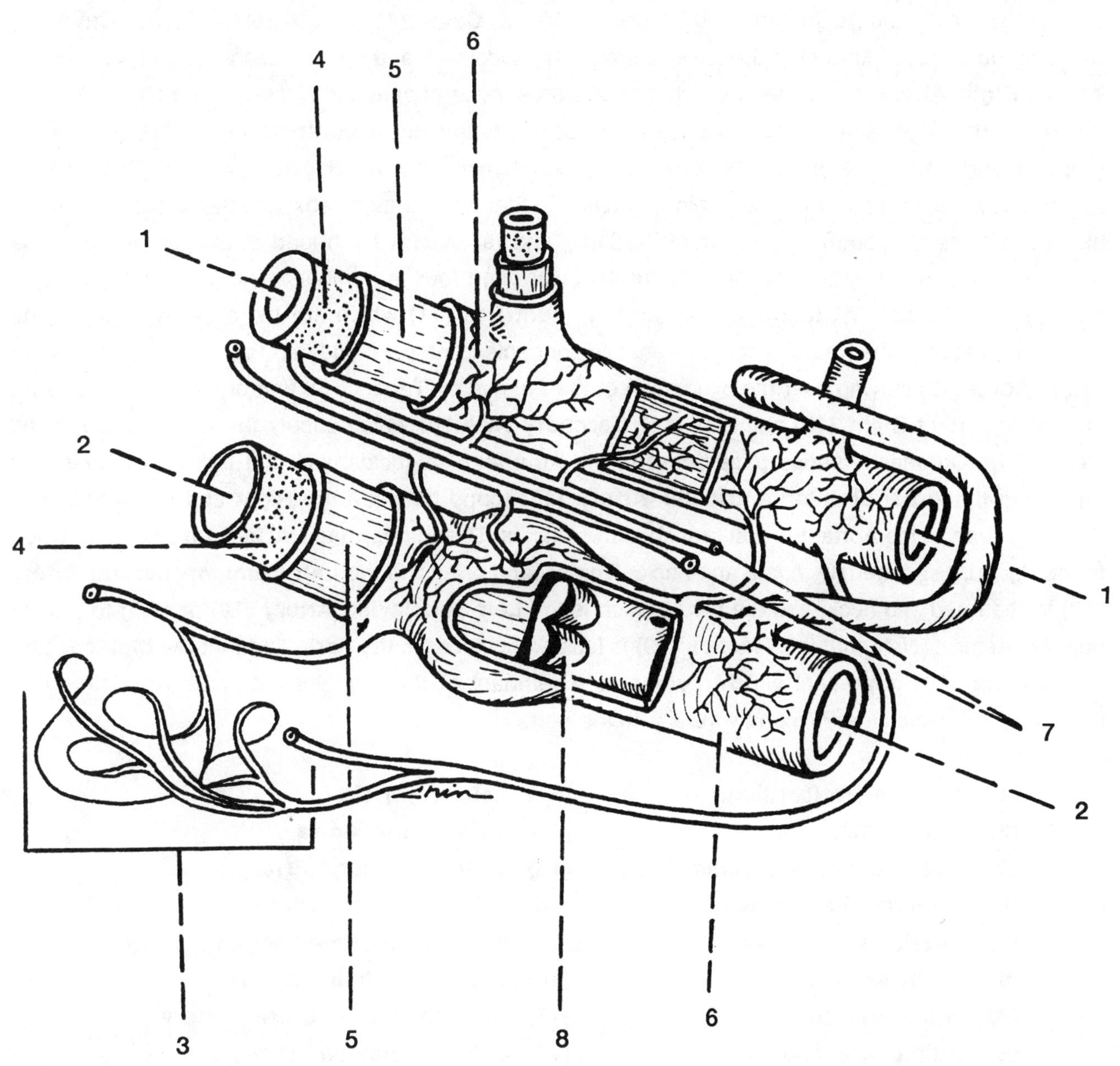

CS-2

Human Heart

(Anterior View)

The Heart is surrounded and protected by a dense connective tissue sac known as the **Fibrous Pericardium (1)** with the **Diaphragm (2)** situated below. Note that the Fibrous Pericardium has been opened to reveal the structure of the Heart and the associated Great Vessels. Venous blood from the upper extremities, head, and neck is returned to the Heart by the Right and Left **Brachiocephalic Veins (3)** which merge to form the **Superior Vena Cava (4)**. The **Inferior Vena Cava (5)** conveys the venous blood from the lower extremities, abdomen, and chest. Each Vena Cava empties into the **Right Atrium** of the Heart which includes an ear-like appendage known as the **Right Auricle (6)**. From the Right Atrium, the venous blood passes into the **Right Ventricle (7)** which contracts to propel blood into the **Pulmonary Trunk (8)** and hence into the **Right** and **Left Pulmonary Arteries (9)**. Realize that these are unique arteries in that they contain deoxygenated blood. Once the blood has passed through the tissues of the Lung and received a fresh load of oxygen, the blood is returned to the Heart by the **Pulmonary Veins (10)**, often four in number. This freshly oxygenated blood enters the **Left Atrium** and then the **Left Ventricle (11)** which are more easily seen on the posterior surface of the Heart. The Left Ventricle contracts to propel blood into the **Aorta** with the **Arch of the Aorta (12)** curving over the bifurcation of the Pulmonary Trunk. The first branches off the Aorta are the important **Right (13)** and **Left (14) Coronary Arteries** which supply the Myocardium of the Heart. The Coronary Arteries pass between the Atria and Ventricles, with the Left Coronary Artery producing the **Anterior Interventricular Artery (15)** found between the Right and Left Ventricles. From the Arch of the Aorta, the first branch is the singular **Brachiocephalic Artery (16)** sending blood to the right upper extremity, head, and neck. The second branch is the **Left Common Carotid Artery (17)** to the head and neck, and the third branch is the **Left Subclavian Artery (18)** passing to the left upper extremity. Note that the **Trachea (19)** is located posterior to the Aorta and that the **Ligamentum Arteriosum (20)** is present which represents a remnant of the embryonic Ductus Arteriosus that previously connected the Pulmonary trunk with the Aorta.

(1) Fibrous Pericardium
(2) Diaphragm
(3) Brachiocephalic Veins
(4) Superior Vena Cava
(5) Inferior Vena Cava
(6) Right Auricle
(7) Right Ventricle
(8) Pulmonary Trunk
(9) Pulmonary Arteries
(10) Pulmonary Veins
(11) Left Ventricle
(12) Arch of the Aorta
(13) Right Coronary Artery
(14) Left Coronary Artery
(15) Anterior Interventricular Artery
(16) Brachiocephalic Artery
(17) Left Common Carotid Artery
(18) Left Subclavian Artery
(19) Trachea
(20) Ligamentum Arteriosum

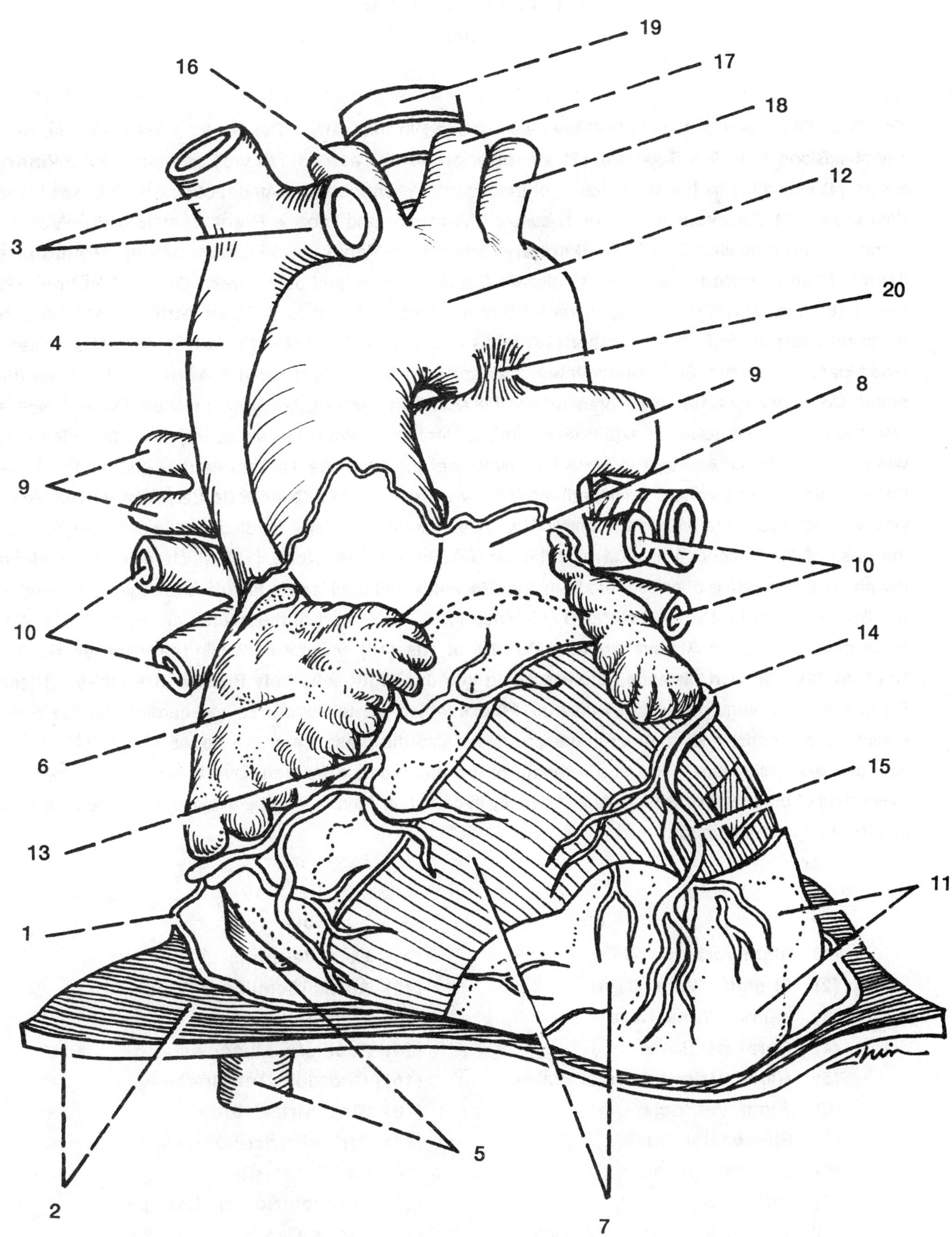

CS-3

Human Heart

(Interior View)

The Four Chambers of the Heart have been opened in this view. The **Right Atrium (1)** receives Venous Blood from the **Superior (2)** and **Inferior Vena Cava (3)**, as well as from the **Coronary Sinus (4)** which drains the Myocardium of the Heart. When the Right Atrium contracts, it sends blood past the **Right Atrioventricular** (or Tricuspid) **Valve (5)** and into the **Right Ventricle (6)**. Venous blood is then propelled past the **Pulmonary Semilunar Valve** (not pictured), into the **Pulmonary Trunk (7)** and carried to the Lungs to discard Carbon Dioxide and obtain fresh Oxygen. **Pulmonary Veins (8)** (typically four in number) return this oxygenated blood to the **Left Atrium (9)**. The **Left Atrioventricular** (or Bicuspid) **Valve (10)** leads to the **Left Ventricle (11)**, which contracts to send blood past the **Aortic Semilunar Valve (12)** and on into the **Arch of the Aorta (13)**. Notice the paired **Coronary Arteries** which branch from the Ascending Aorta just beyond the Aortic Valve. Notice also the presence of rounded extensions from the Ventricular Walls known as **Papillary Muscles (14)** which connect to the Atrioventricular Valves by way of the **Chordae Tendineae (15)** and act to control the opening of these Valves. Included in this view is the important **Conduction System of the Heart** which determines the rate of Heart Contraction. This System of modified cardiac muscle cells begins near the base of the Superior Vena Cava at the **Sino-Atrial (or S-A) Node (16)** which initiates the Heart rhythm and is therefore often referred to as the "Pacemaker of the Heart". Impulses are then conveyed to the **Atrioventricular (or A-V) Node (17)** located in the lower part of the Interatrial Septum. The A-V Node gives rise to the **Atrioventricular Bundle of His (18)** which runs within the **Interventricular Septum (19)** a short distance before splitting into the **Right and Left Bundle Branches**. These Branches form numerous small **Purkinje Fibers** which act to depolarize the cardiac muscle cells, causing the Ventricles to contract in unison. Note also the presence of the **Fossa Ovalis (20)** in the wall of the Right Atrium. This is a remnant of the embryonic **Foramen Ovalis** which permitted the movement of blood from the Right Atrium directly into the Left Atrium, thereby avoiding passage to the still inactive Lungs of the unborn child.

(1) Right Atrium
(2) Superior Vena Cava
(3) Inferior Vena Cava
(4) Coronary Sinus
(5) Right Atrioventricular Valve
(6) Right Ventricle
(7) Pulmonary Trunk
(8) Pulmonary Veins
(9) Left Atrium
(10) Left Atrioventricular Valve
(11) Left Ventricle
(12) Aortic Semilunar Valve
(13) Arch of the Aorta
(14) Papillary Muscles
(15) Chordae Tendineae
(16) Sino-Atrial Node
(17) Atrioventricular Node
(18) Bundle of His
(19) Interventricular Septum
(20) Fossa Ovalis

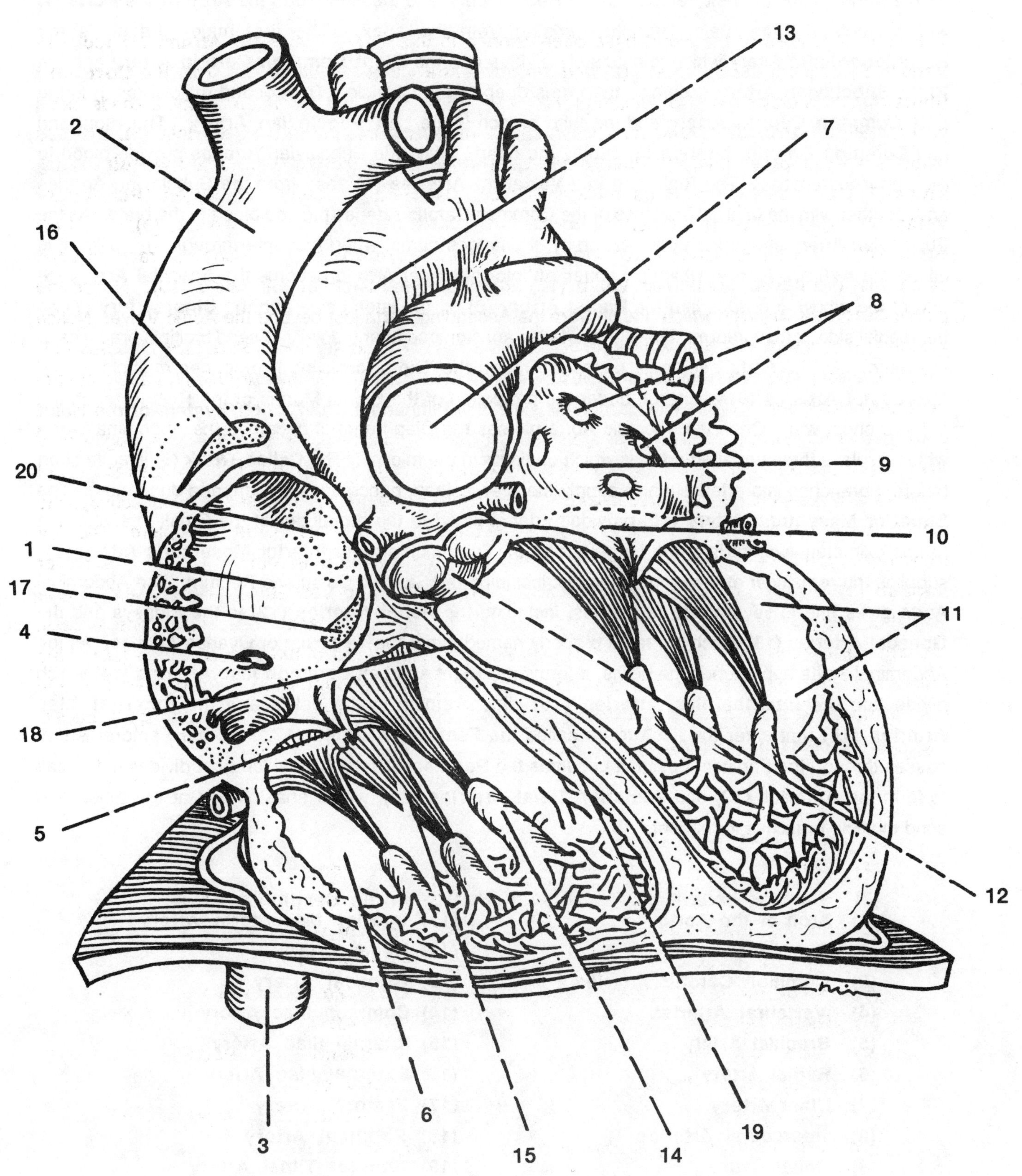

CS-4

Primary Systemic Arteries

As the Aorta leaves the Left Ventricle of the Heart, it curves to the left to form the **Arch of the Aorta (1)** which displays three major branches arising from the apex. The first major branch is the **Brachiocephalic** Artery which immediately divides into the **Right Common Carotid Artery** and the **Right Subclavian Artery (2)** which then runs deep to the Clavicle. The second major branch is the **Left Common Carotid Artery** and the third branch is the **Left Subclavian Artery**. The Right and Left **Common Carotid Arteries (3)** supply the head, while the Subclavian Arteries provide blood to the upper extremities. The Right and Left **Vertebral Arteries (4)** arise from the Subclavian Arteries and together with the Internal branches of the Common Carotid Arteries provide blood to the brain. As the Subclavian Artery enters the upper extremity, it changes name: when it course through the Axilla, it is called the **Axillary Artery**, which continues on into the upper Arm to become the **Brachial Artery (5)** and at the elbow, it divides into the **Radial Artery (6)** on the lateral side and the **Ulnar Artery (7)** on the medial side of the forearm. These two arteries anastomose near the wrist to send multiple branches to the hand. As the Aorta begins its descent through the Thoracic cavity, it produces multiple paired **Intercostal Arteries (8)** which run under the protection of the Costal Margin of the Ribs to reach the anterior chest wall. Once the Thoracic Aorta pierces the Diaphragm, it becomes the Abdominal Aorta which displays three unpaired arteries which originate in the midline: The **Celiac Trunk (9)** is quite short before it branches into arteries which supply the liver, spleen, pancreas, stomach, and duodenum. The **Superior Mesenteric Artery (10)** provides arterial blood to the remainder of the Small Intestine and to approximately two-thirds of the Large Intestine. The much smaller **Inferior Mesenteric Artery (11)** supplies the remainder of the Large Intestine, including the Sigmoid Colon and Rectum. The Abdominal Aorta gives rise to several paired arteries, including the **Renal Arteries (12)** to the Kidneys and the **Gonadal Arteries (13)** which are more properly named either the Testicular or Ovarian Arteries. As the Abdominal Aorta approaches the pelvis, it forms the Right and Left **Common Iliac Arteries (14)** which divide into the **Internal Iliac Arteries (15)** which remain in the pelvis and the **External Iliac Arteries (16)** that enter the thigh to be called the **Femoral Arteries (17)**. As the Femoral Artery passes through the Popliteal Fossa, it becomes the **Popliteal Artery (18)** which then divides in the calf to form the **Anterior (19)** and **Posterior Tibial Arteries (20)**, which anastomose at the ankle and send multiple branches to the foot.

(1) Arch of the Aorta
(2) Subclavian Artery
(3) Common Carotid Arteries
(4) Vertebral Arteries
(5) Brachial Artery
(6) Radial Artery
(7) Ulnar Artery
(8) Intercostal Arteries
(9) Celiac Trunk
(10) Superior Mesenteric Artery
(11) Inferior Mesenteric Artery
(12) Renal Artery
(13) Gonadal Artery
(14) Common Iliac Artery
(15) Internal Iliac Artery
(16) External Iliac Artery
(17) Femoral Artery
(18) Popliteal Artery
(19) Anterior Tibial Artery
(20) Posterior Tibial Artery

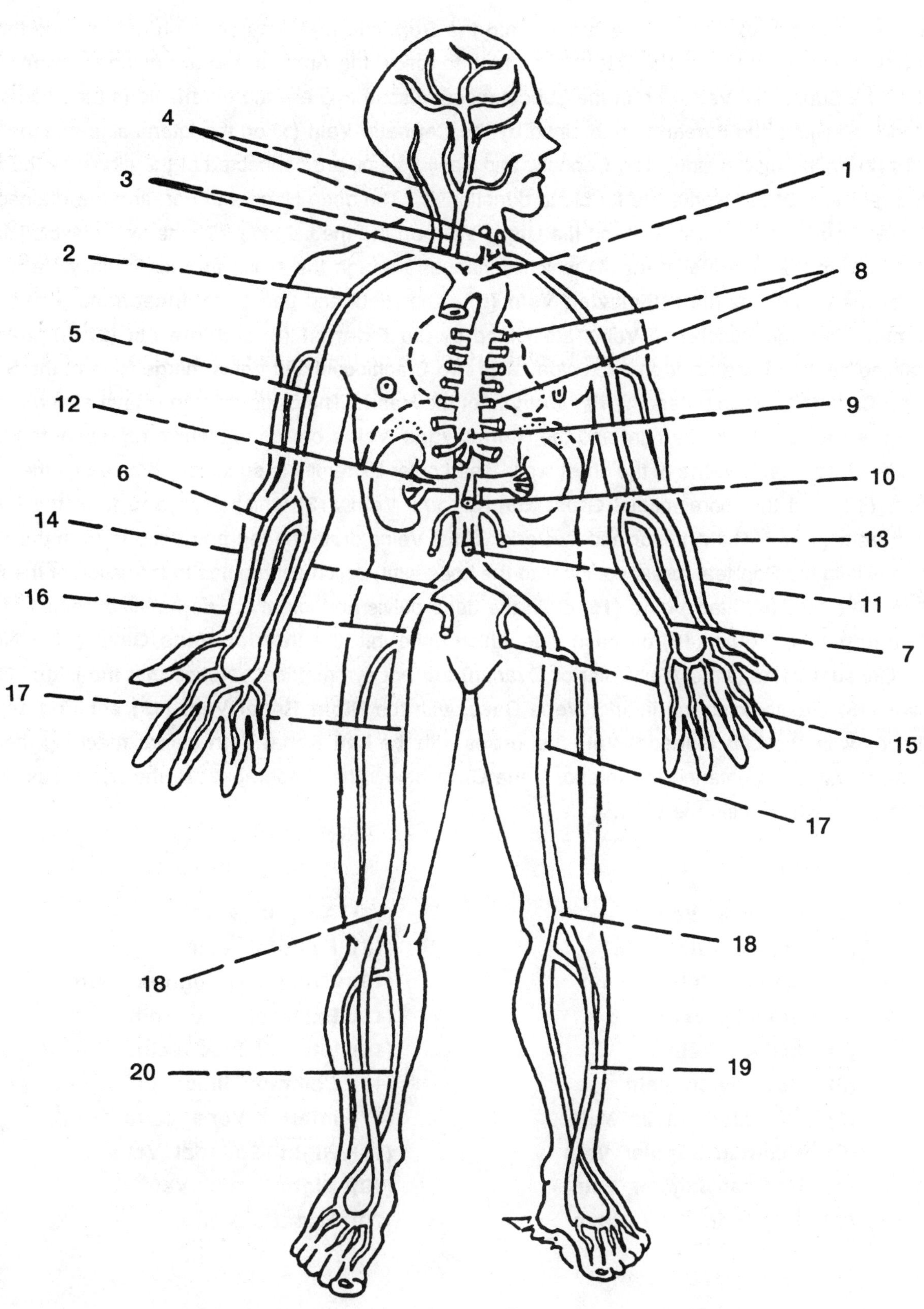
4
3
1
2
8
5
12
9
6
10
14
13
11
16
7
17
15
17
18
18
20
19

Primary Systemic Veins

The Veins of the extremities are divided into the **Superficial Veins** seen on the left and the **Deep Veins** seen on the right. The Deep Veins tend to follow the Arteries and are given the same names, while the Superficial Veins run in the Subcutaneous tissue and are more variable in their course. The superficial Hand and Forearm are drained by the **Cephalic Vein (1)** on the lateral side and the **Basilic Vein (2)** on the medial side. The Cephalic and Basilic Veins are connected at the elbow by the **Median Cubital Vein (3)** (a favorite site for blood donations!). The deep Hand and Forearm are drained by the **Radial Vein** on the lateral side and the **Ulnar Vein** on the medial side. These two vessels join in the Arm to form the **Brachial Vein (4)** which continues through the Axilla as the **Axillary Vein (5)** and under the clavicle as the **Subclavian Vein (6)**. The Right and Left **Brachiocephalic Veins (7)** are formed when the Subclavian Veins are joined by the **External (8)** and **Internal (9) Jugular Veins** draining the head and neck on each side. The two Brachiocephalic Veins merge to form the **Superior Vena Cava (10)** which enters the Right Atrium of the Heart. The Superior Vena Cava also receives the **Azygos Veins (11)** on the right and the **Hemiazygos Veins** on the left which represent the union of multiple **Intercostal Veins** of the chest wall. The Lower Extremity is principally drained by the **Femoral Vein (12)** and the more medial **Great Saphenous Vein (13)** which merge to form the **External Iliac Vein (14)**. The Anterior and Posterior Tibial Veins drain the leg and unite to from the Popliteal Vein (within the Popliteal fossa, posterior to the knee joint) which contributes to formation of the Femoral Vein. The **Internal Iliac Veins (15)** drain the deep pelvis and join the External Iliac Veins to form the **Common Iliac Veins (16)** on each side which unite into the **Inferior Vena Cava (17)**. Note that the Gonadal Veins (either Testicular or Ovarian) are not symmetrical. Notice that the **Right Gonadal Vein (18)** directly joins the Inferior Vena Cava, with the **Right Renal Vein (19)** entering separately above, while the **Left Gonadal Vein** first unites with the **Left Renal Vein** before reaching the Inferior Vena Cava. Also entering the Inferior Vena Cava below the Diaphragm are the numerous **Hepatic Veins (20)** which drain the Liver.

(1) Cephalic Vein
(2) Basilic Vein
(3) Median Cubital Vein
(4) Brachial Vein
(5) Axillary Vein
(6) Subclavian Vein
(7) Brachiocephalic Vein
(8) External Jugular Vein
(9) Internal Jugular Vein
(10) Superior Vena Cava
(11) Azygos Vein
(12) Femoral Vein
(13) Great Saphenous Vein
(14) External Iliac Vein
(15) Internal Iliac Vein
(16) Common Iliac Vein
(17) Inferior Vena Cava
(18) Right Gonadal Vein
(19) Right Renal Vein
(20) Hepatic Veins

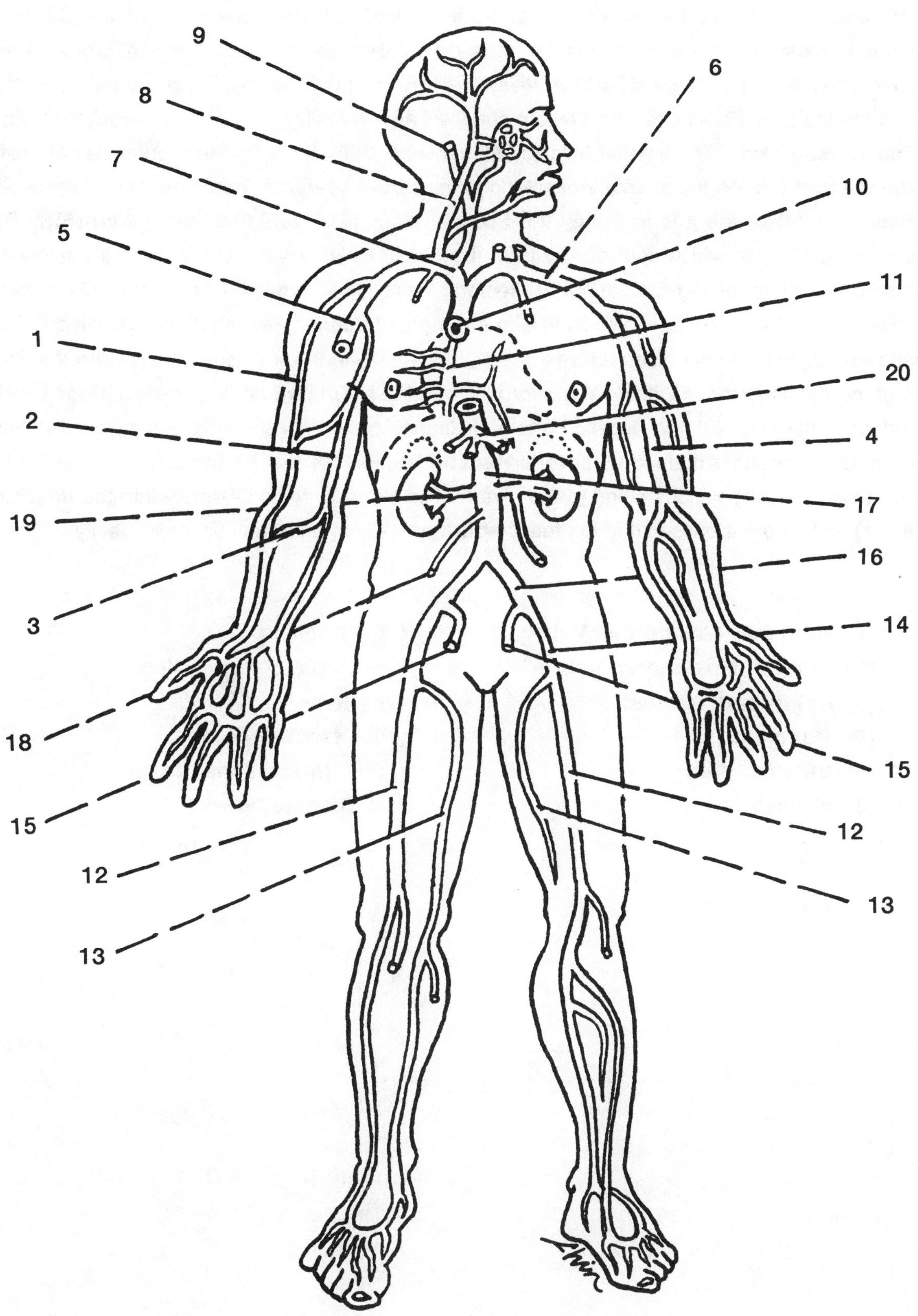
9
6
8
7
10
5
11
1
20
2
4
17
19
16
3
14
18
15
15
12
12
13
13

Hepatic Portal Veins

The **Hepatic Portal System of Veins** serves to connect the capillary beds of the Stomach and Intestines to those within the Liver, thereby transporting absorbed nutrients for final processing by the Liver Hepatocytes. The large **Superior Mesenteric Vein (1)** is formed from a multitude of smaller veins which drain the Small Intestine, Cecum and the Large Intestine (or Colon) to about the midpoint of the Transverse Colon. The smaller **Inferior Mesenteric Vein (2)** drains the remainder of the **Colon** and the **Rectum (3)**. As the Inferior Mesenteric Vein courses upward toward the Liver, it passes behind the **Pancreas (4)** where it is joined by the **Splenic Vein (5)**. Note that the **Spleen (6)** includes a great many tributaries which coalesce to form the single Splenic Vein which passes posterior to the **Stomach (7)** before joining the Inferior Mesenteric Vein. The **Hepatic Portal Vein (8)** is formed by the union of the Superior Mesenteric Vein with the merged Inferior Mesenteric and Splenic Veins. The Hepatic Portal Vein usually also receives directly the Right and Left Gastric Veins from the Stomach, although occasionally these Gastric Veins form interesting alternative anastomoses. Upon penetrating the surface of the **Liver (9)**, the Hepatic Portal Vein immediately divides into a great many branches. The Liver tissue is ultimately drained by several **Hepatic Veins (10)** (not to be confused with the Hepatic Portal Veins that bring venous blood to the Liver). The Hepatic Veins all empty into the **Inferior Vena Cava (11)** just before it passes through the **Diaphragm (12)** to enter the Thoracic Cavity.

(1) Superior Mesenteric Vein
(2) Inferior Mesenteric Vein
(3) Colon and Rectum
(4) Pancreas
(5) Splenic Vein
(6) Spleen
(7) Stomach
(8) Hepatic Portal Vein
(9) Liver
(10) Hepatic Veins
(11) Inferior Vena Cava
(12) Diaphragm

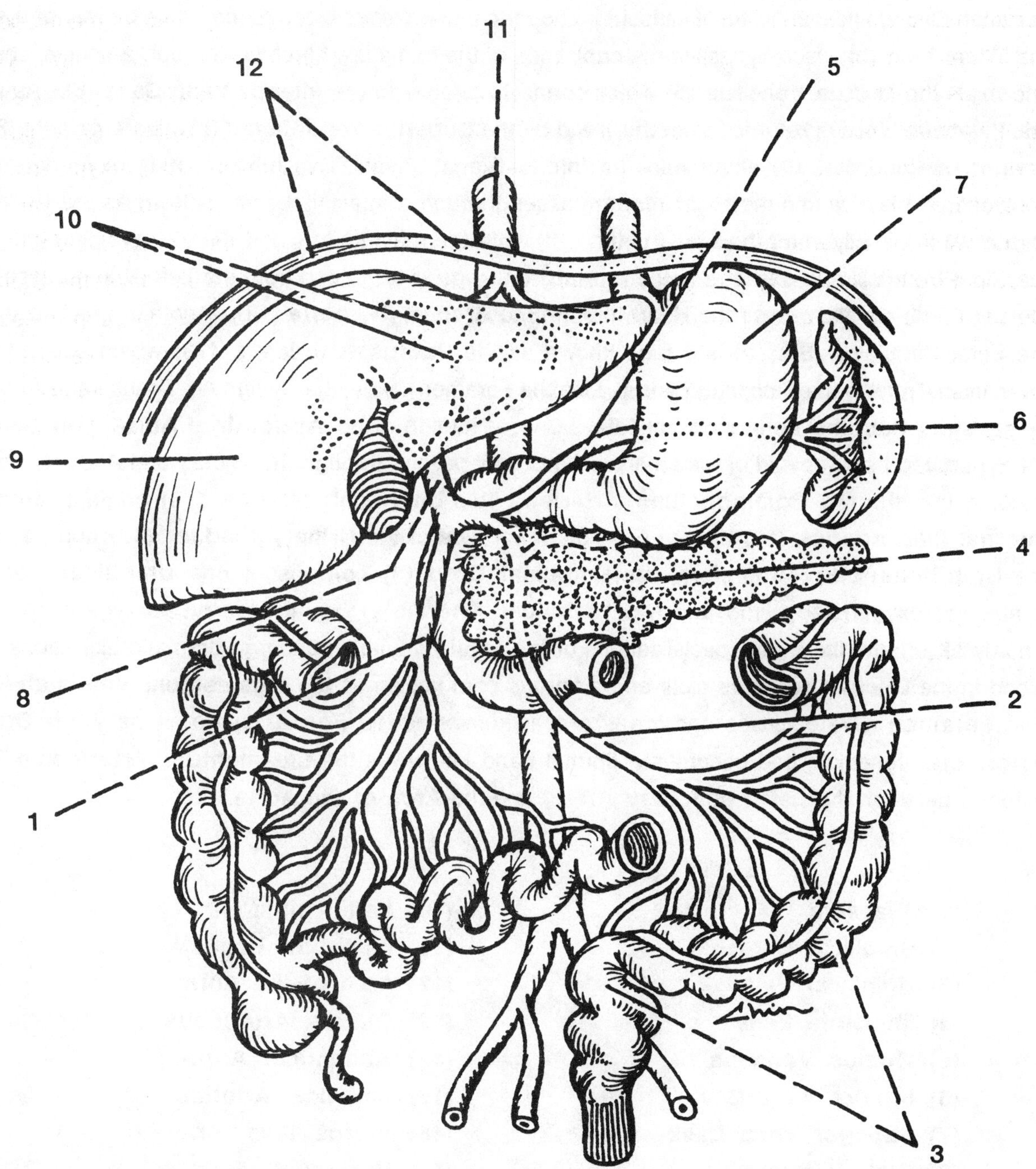

CS-7

Fetal Circulation

The fetus receives oxygen and nutrients from the **Placenta (1)** through the **Umbilical Cord (2)** and also returns waste products including Carbon Dioxide to the Placenta. Three specializations exist in the circulatory system of the unborn child which cease to function after birth: the Ductus Venosus, the Foramen Ovale, and the Ductus Arteriosus. The Placenta provides blood to the Fetus by way of a single **Umbilical Vein (3)** which bypasses the capillaries of the immature **Liver (4)** through a special channel known as the **Ductus Venosus (5)** which connects directly to the **Inferior Vena Cava (6)**. As in the adult, venous blood is returned from the Head by the **Superior Vena Cava (7)** which enters the **Right Atrium of the Heart (8)** along with the Inferior Vena Cava. This mixture of Deoxygenated and Oxygenated blood within the Right Atrium passes through a special opening known as the **Foramen Ovale (9)** to directly enter the **Left Atrium (10)**, with the majority of blood thereby avoiding entry into the Right Ventricle and passage to the **Immature Lungs (11)**. Blood from the Left Atrium passes into the Left Ventricle and leaves the Heart through the **Arch of the Aorta (12)**. Another specialization of the Fetal Circulatory System is a tube known as the **Ductus Arteriosus (13)** which connects the Pulmonary Trunk to the Arch of the Aorta. Like the Foramen Ovale, the Ductus Arteriosus helps to shuttle blood away from the non-functional fetal Lung. Note that the **Abdominal Aorta (14)** contains deoxygenated blood devoid of maternal nutrients as it passes through the Thorax and Abdomen before dividing into the two **Common Iliac Arteries**. Two **Umbilical Arteries (15)** originate from the **Internal Iliac Arteries (16)** to then pass on either side of the **Urinary Bladder (17)** and penetrate the **Umbilicus (18)**. Therefore, the **Umbilical Cord (2)** consists of one **Umbilical Vein (3)** containing oxygenated blood and two **Umbilical Arteries (15)** which carry deoxygenated blood. Shortly after birth, the three specializations of the Fetal Circulation undergo dramatic alterations. The blood in the **Ductus Venosus** clots and a fibrous cord known as the **Ligamentum Venosum** forms. The **Foramen Ovale** closes and forms a scar known as the **Fossa Ovale**. Finally, the **Ductus Arteriosus** constricts and becomes a fibrous band known as the **Ligamentum Arteriosum** which extends between the **Left Pulmonary Artery** and the **Arch of the Aorta**.

(1) Placenta
(2) Umbilical Cord
(3) Umbilical Vein
(4) Immature Liver
(5) Ductus Venosus
(6) Inferior Vena Cava
(7) Superior Vena Cava
(8) Right Atrium
(9) Foramen Ovale
(10) Left Atrium
(11) Immature Lungs
(12) Arch of the Aorta
(13) Ductus Arteriosus
(14) Abdominal Aorta
(15) Umbilical Arteries
(16) Internal Iliac Artery
(17) Urinary Bladder
(18) Umbilicus

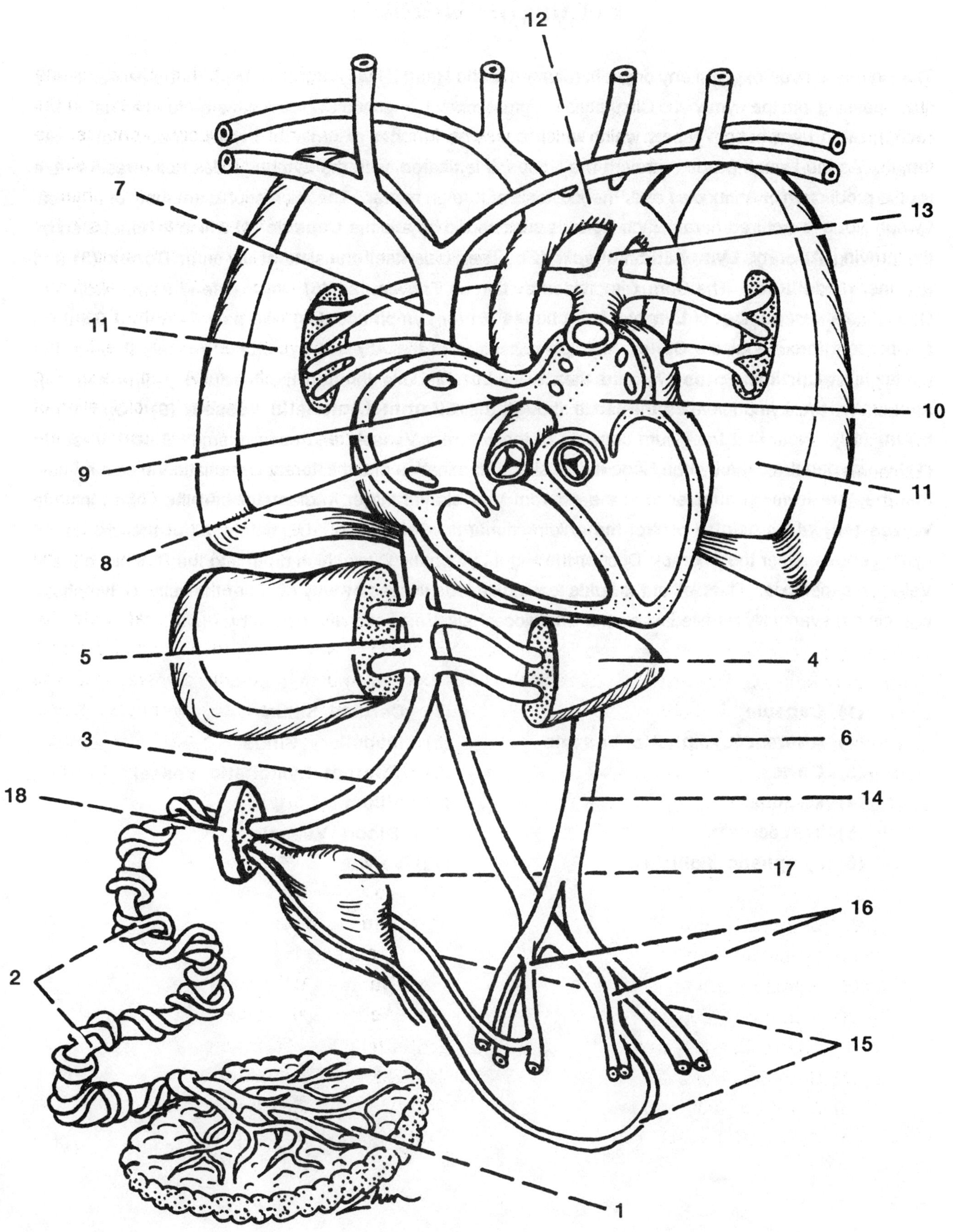
12
7
13
11
10
9
11
8
5
4
3
6
18
14
17
16
2
15
1

Lymph Node

The Arterial Blood Supply to any organ is returned to the Heart by way of the Veins, with the excess tissue fluids passing into the Lymphatic Circulation. A great many Lymphatic Capillaries merge to eventually form recognizable Lymphatic Vessels which include a large number of regional Lymph Nodes along their length. As the Lymph passes through the Nodes, it is filtered, with the Lymph Nodes being responsible for the production of Antibodies and Phagocytosis of foreign matter. The microscopic anatomy of a typical Lymph Node is pictured here. Each Node is surrounded by a loose **Capsule (1)** which is penetrated by the arriving **Afferent Lymphatic Vessels (2)**. The Node itself consists of an outer **Cortex (3)** and an inner **Medulla (4)**. The Cortex is divided by fibrous **Trabeculae (5)** which extend inward from the Capsule. A vast number of **Lymphatic Follicles (6)** (or Lymph nodules) with pale **Germinal Centers** are present throughout the Cortex. Lymph passes from the Afferent Lymphatic Vessels through the Cortex and **Cortical Sinus (7)**, and then on to the Medulla through the **Medullary Sinuses (8)**. Eventually, the Lymph leaves the Node though the **Efferent Lymphatic Vessels (9)** found at the **Hilum (10)**. Note that the Hilum consists of the Efferent Vessels and several small **Blood Vessels (11)** which penetrate the Lymph Node at this site. Examination of the Efferent Lymphatic Vessels reveals that they are fewer in number than the Afferent Vessels yet larger in diameter and they often include **Valves (12)** which help to control the unidirectional flow of Lymph. Ultimately, all of the circulating Lymph enters either the Thoracic Duct or the Right Lymphatic Duct which drain into the Brachiocephalic Veins on either side. Thus, all of the fluids leaving various tissues by way of either the veins or lymphatic vessels are eventually reunited within the Brachiocephalic Veins and returned to the Heart.

(1) Capsule
(2) Afferent Lymphatic Vessels
(3) Cortex
(4) Medulla
(5) Trabecula
(6) Lymphatic Follicle
(7) Cortical Sinus
(8) Medullary Sinus
(9) Efferent Lymphatic Vessel
(10) Hilum
(11) Blood Vessels
(12) Valve

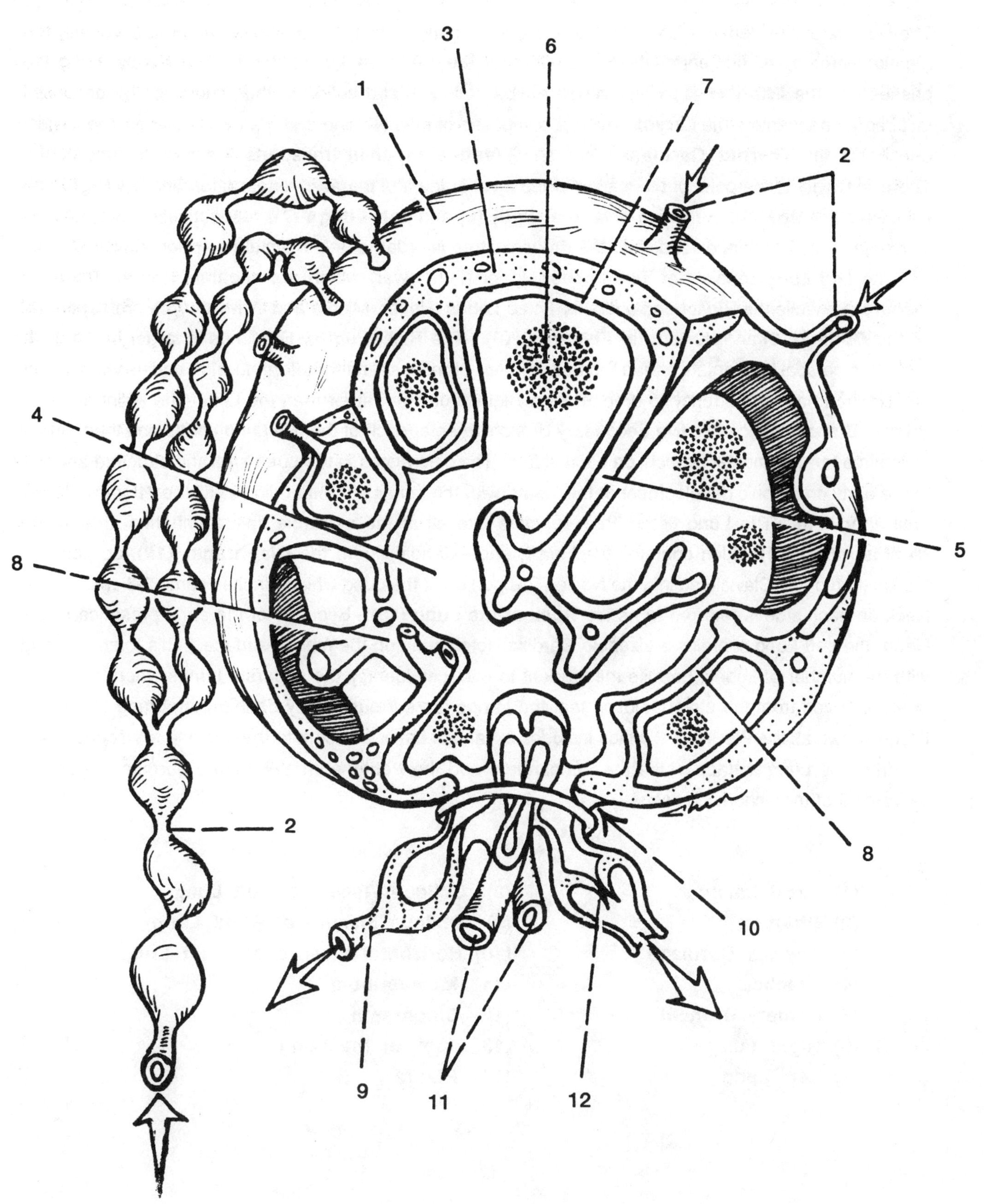

CS-9

Respiratory System
Gross Anatomy

The Respiratory System is responsible for the gas exchange which occurs between the atmosphere and the bloodstream. Air first enters the System through the **Nasal Cavity (1)** (not the Oral Cavity!) and then passes into the **Pharynx (2)** which also transports liquids and solids to the Digestive System. The inspired air next enters the **Larynx** which is composed of nine Hyaline and Elastic Cartilages, the largest one being the **Thyroid Cartilage (3)**. The **Trachea (4)** then transports the gases through the Thoracic Cavity to the point of bifurcation at the level of the fifth thoracic vertebra, forming the **Right** and **Left Primary Bronchi (5)** which enter the **Right (6)** and **Left Lungs (7)**, respectively. The Hyaline Cartilage of the Trachea extends into the Bronchial tree in order to help maintain an open airway. Notice that the Left Lung consists of Two Lobes (Upper and Lower) which are separated by an **Oblique Fissure (8)**, while the Right Lung contains Three Lobes (Upper, Middle and Lower) which are separated by the **Oblique (9)** and **Horizontal Fissures (10)**. The Right Primary Bronchus is larger in diameter and more vertical in orientation than the Left Primary Bronchus, and therefore, foreign bodies which enter the Trachea have a tendency to drop into the Right Bronchus rather than the Left. The Right and Left Primary Bronchi branch into five Secondary Bronchi to enter each of the five Lung Lobes and then further divide into ten Segmental Bronchi on each side. The differences in the Lung size and structure are due to the early embryonic development of the Heart, with the Lungs not expanding to fill the Thoracic Cavity until after birth. The Lung tissue then reaches into all of the available space, inflating around the **Mediastinum (11)** which contains the Heart, and extending from the **Diaphragm (12)**, up past the medial third of the Clavicle on into the Neck. This portion of the Lung which reaches some 2-3 cm. into the Neck on each side is referred to as the **Apex of the Lung (13)**. Because of the early presence of the Heart, the Left Lung contains a sizeable Cardiac Notch seen on the medial surface. This feature, along with the number of Lobes, permits the student to correctly identify the separate Lungs when they have been removed from the chest. Notice that the Lungs are surrounded by double membrane of Fibrous tissue known as the **Pleura (14)** which includes a Parietal Layer adjacent to the ribs and a Visceral Layer covering the Lung surface. Like the Pericardium and the Peritoneum, the Pleura permits frictionless movement of the underlying tissue.

(1) Nasal Cavity
(2) Pharynx
(3) Thyroid Cartilage
(4) Trachea
(5) Primary Bronchi
(6) Right Lung
(7) Left Lung
(8) Oblique Fissure of Left Lung
(9) Oblique Fissure of Right Lung
(10) Horizontal Fissure of Right Lung
(11) Mediastinum
(12) Diaphragm
(13) Apex of the Lung
(14) Pleura

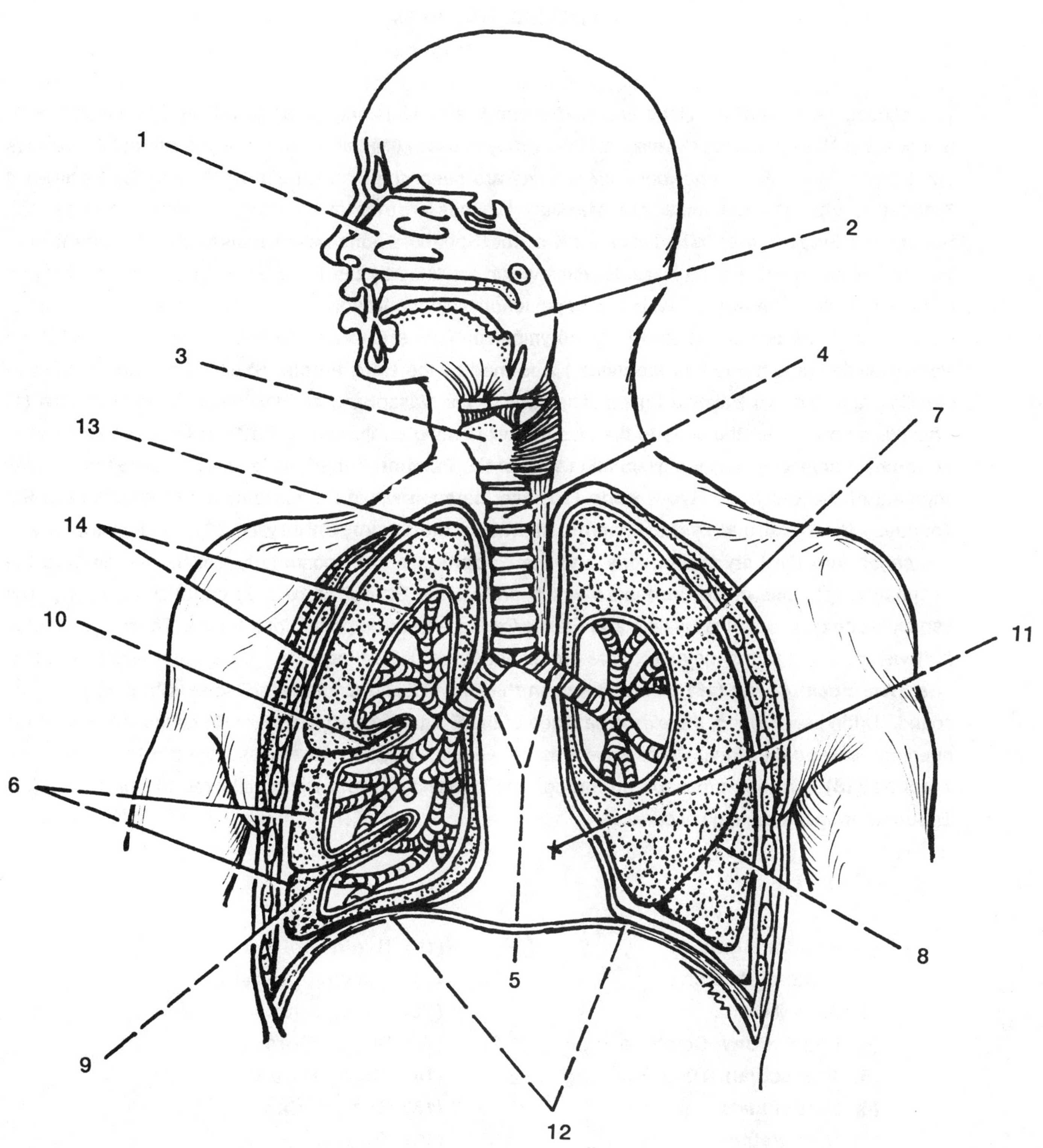
1
2
3
4
13
7
14
10
11
6
8
5
9
12

Head and Neck
(Midsagittal View)

This Midsagittal View of the Head and Neck permits an understanding of the relations between the Air Sinuses, the Nasal and Oral Cavities, and the Pharynx which ultimately divides below into the Esophagus and Larynx. Two of the Four bony air sinuses are seen: the **Frontal Sinus (1)** and the **Sphenoid Sinus (2)**, while the Ethmoidal and Maxillary Sinuses are not visible here. Notice that above the Sphenoidal Sinus is the **Sella Turcica (3)** of the Sphenoid bone which houses the Pituitary Gland. The air sinuses drain into the Nasal Cavities which are lined by the three bony **Nasal Conchae (4)** on each lateral wall. The Nasal Cavities open posteriorly into the **Nasopharynx** which receives the bilateral openings of the **Eustachian** (or Pharyngotympanic) **Tubes (5)** from the middle ear. The floor of the Nasal Cavities (and the roof of the mouth) is formed by the **Hard Palate (6)** which consists of the two Maxillary and the two Palatine bones. The floor of the Nasopharynx is the muscular **Soft Palate (7)** which flips up to cover the entry to the Nasal Cavity during swallowing. The **Oropharynx** includes the **Tongue (8)** anteriorly and the lymphatic tissue of the **Palatine Tonsil (9)** laterally. Notice that a small segment of the U-shaped **Hyoid bone (10)** may be seen within the surrounding musculature of the Tongue. The distal end of the Pharynx, known as the **Laryngopharynx (11)**, marks the site of bifurcation into the Larynx anteriorly and the **Esophagus (12)** posteriorly. Note that deep to the Esophagus, portions of the Vertebrae which protect the **Spinal Cord (13)** can also be seen. The Larynx is composed of nine Hyaline and Elastic Cartilages, with a portion of the large **Thyroid Cartilage (14)** visible, along with segments of the prominent **Cricoid Cartilage (15)** located immediately below. The **True Vocal Cords (16)** extend between the Thyroid and Cricoid Cartilages, vibrating to create sound. During swallowing, the elastic cartilage of the **Epiglottis (17)** flips down to cover and protect the opening between the Vocal Cords known as the Glottis. Inspired Air moves from the Larynx into the **Trachea (18)** and then proceeds on through the remainder of the Bronchial Tree to reach the Lung Tissue where gas exchange occurs.

(1) Frontal Sinus
(2) Sphenoid Sinus
(3) Sella Turcica
(4) Nasal Bony Conchae
(5) Eustachian Tube opening
(6) Hard Palate
(7) Soft Palate
(8) Tongue
(9) Palatine Tonsil
(10) Hyoid Bone
(11) Laryngopharynx
(12) Esophagus
(13) Spinal Cord
(14) Thyroid Cartilage
(15) Cricoid Cartilage
(16) True Vocal Cords
(17) Epiglottis
(18) Trachea

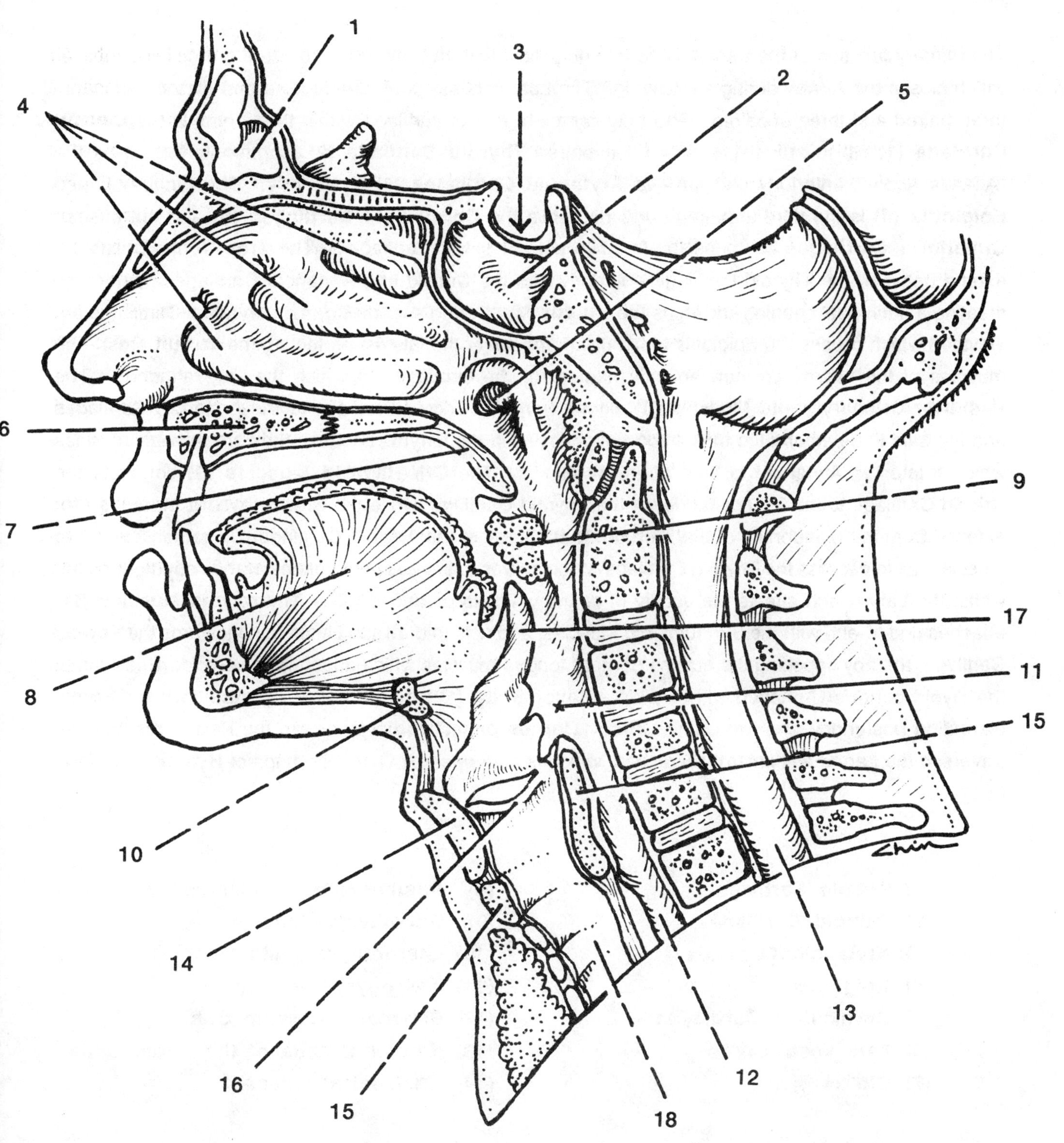

RS-2

Larynx
(Side View)

The primary function of the Larynx is not sound production, but rather to provide a protective sphincter which closes the Airway during swallowing. The Larynx consists of nine hyaline and elastic cartilages, three paired and three unpaired. The only complete ring of cartilage within the Larynx is the **Cricoid Cartilage (1)** which articulates with the unpaired **Thyroid Cartilage (2)** (commonly known as the "Adam's Apple") anteriorly and the two **Arytenoid Cartilages (3)** posteriorly. The tongue-shaped **Epiglottis (4)** is the third unpaired cartilage, while the **Corniculate Cartilages (5)** and the unseen **Cuneiform Cartilages** comprise the last two sets of paired cartilages. The **True Vocal Cords (6)** extend between the Thyroid Cartilage anteriorly and the Cricoid and Arytenoid Cartilages posteriorly, creating a laryngeal opening known as the **Glottis (7)**. During swallowing, the Laryngeal Cartilages are elevated which causes the Epiglottis to flip down and cover the Glottis, protecting the Vocal Cords. The muscles of the Larynx are conveniently named for the cartilages to which they are attached. The **Posterior Cricoarytenoid Muscles (8)** extend from the Cricoid Cartilage to the Arytenoid Cartilages and are said to be among the most important muscles in the Larynx because they are the only muscles which rotate the Arytenoid to then open the Glottis. The **Cricothyroid Muscles (9)** connects the Cricoid Cartilage to the inferior border of the Thyroid Cartilage, while the **Sternothyroid Muscles (10)** extends from the Manubrium of the Sternum to the body of the Thyroid Cartilage. Together, these two muscles act to depress the Thyroid Cartilage which serves to increase the volume of the sound generated within the Larynx and affects the quality of sound which is produced. The **Thyrohyoid Muscles (11)** attaches end-to-end with the Sternothyroid Muscles, and acts to draw the Hyoid bone towards the Thyroid Cartilage, thereby opposing the actions of other tongue muscles which attach to the Hyoid. Realize that the Hyoid bone is U-shaped with a wide anterior **Body (12)**, two **Greater Cornua (13)** (or Horns) extending posteriorly and two small **Lesser Cornua** projecting upward from the Body. Air that has traversed the Larynx will enter the Trachea, which is held open by C-shaped rings of **Hyaline Cartilage (14)**.

(1) Cricoid Cartilage
(2) Thyroid Cartilage
(3) Arytenoid Cartilage
(4) Epiglottis
(5) Corniculate Cartilages
(6) True Vocal Cords
(7) Glottis
(8) Posterior Cricoarytenoid Muscles
(9) Cricothyroid Muscle
(10) Sternothyroid Muscle
(11) Thyrohyoid Muscle
(12) Body of the Hyoid Bone
(13) Greater Cornua of the Hyoid Bone
(14) Tracheal Cartilages

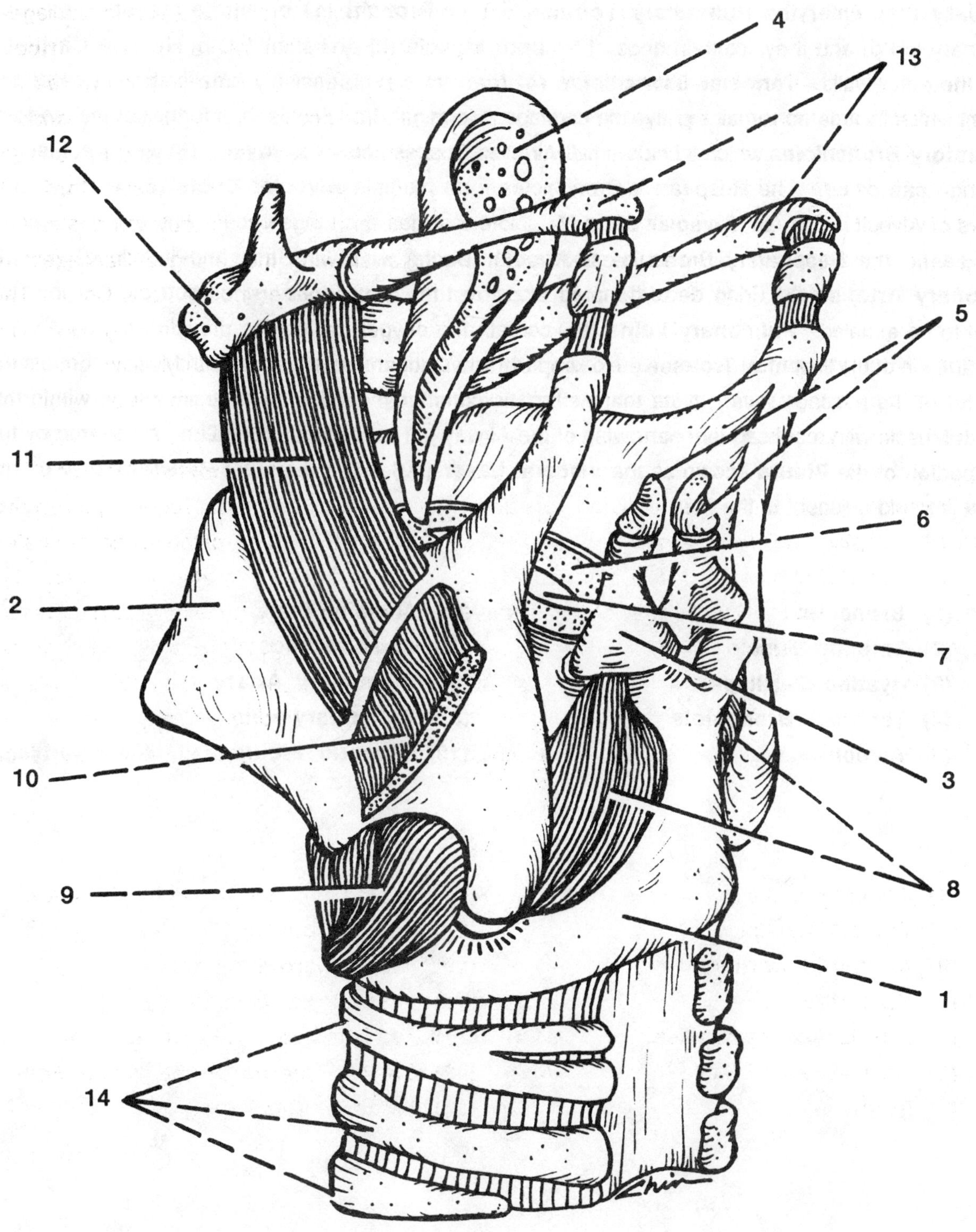
4
13
12
5
11
6
2
7
3
10
8
9
1
14

Pulmonary Lobule
(Histology)

As Air travels through the **Bronchial Tree**, the branches become progressively smaller and smaller until eventually they enter the **Pulmonary Lobules.** The **Bronchi (1)** penetrate the apex of each pulmonary lobule and they contain rings of **Smooth Muscle (2)** and small bits of **Hyaline Cartilage (3)** in the outer walls. **Terminal Bronchioles (4)** form from progressively smaller branches of the Bronchi which include no remaining Hyaline cartilage. Terminal Bronchioles then further branch to form **Respiratory Bronchioles** which include small Air-filled Spaces known as **Alveoli (5)** where actual gas exchange can occur. The Respiratory Bronchioles form multiple **Alveolar Ducts (6)** which lead to clusters of Alveoli known as **Alveolar Sacs (7)**. Note that gas exchange occurs wherever the Alveoli are present: the Respiratory Bronchioles, Alveolar Ducts, Alveolar Sacs and individual Alveoli. **Pulmonary Arteries (8)** bring deoxygenated blood to the lung, depositing carbon dioxide into the Alveoli to be exhaled. **Pulmonary Veins (9)** contain the oxygen molecules provided by the Alveoli. The Cartilage of the Bronchial Tree serves to keep the Airway open, while the Smooth Muscle controls the diameter of the passageways. (Note that Asthma is commonly caused by excessive tone within this Smooth Muscle with a subsequent narrowing of the Airway.) The Surface of the Lung is covered by the inner portion of the **Pleura** known as the **Visceral Layer (10)**, while the outer **Parietal Layer of the Pleura** is found adjacent to the ribs.

(1) Bronchus
(2) Smooth Muscle
(3) Hyaline Cartilage
(4) Terminal Bronchiole
(5) Alveoli
(6) Alveolar Duct
(7) Alveolar Sacs
(8) Pulmonary Artery
(9) Pulmonary Vein
(10) Visceral Pleura

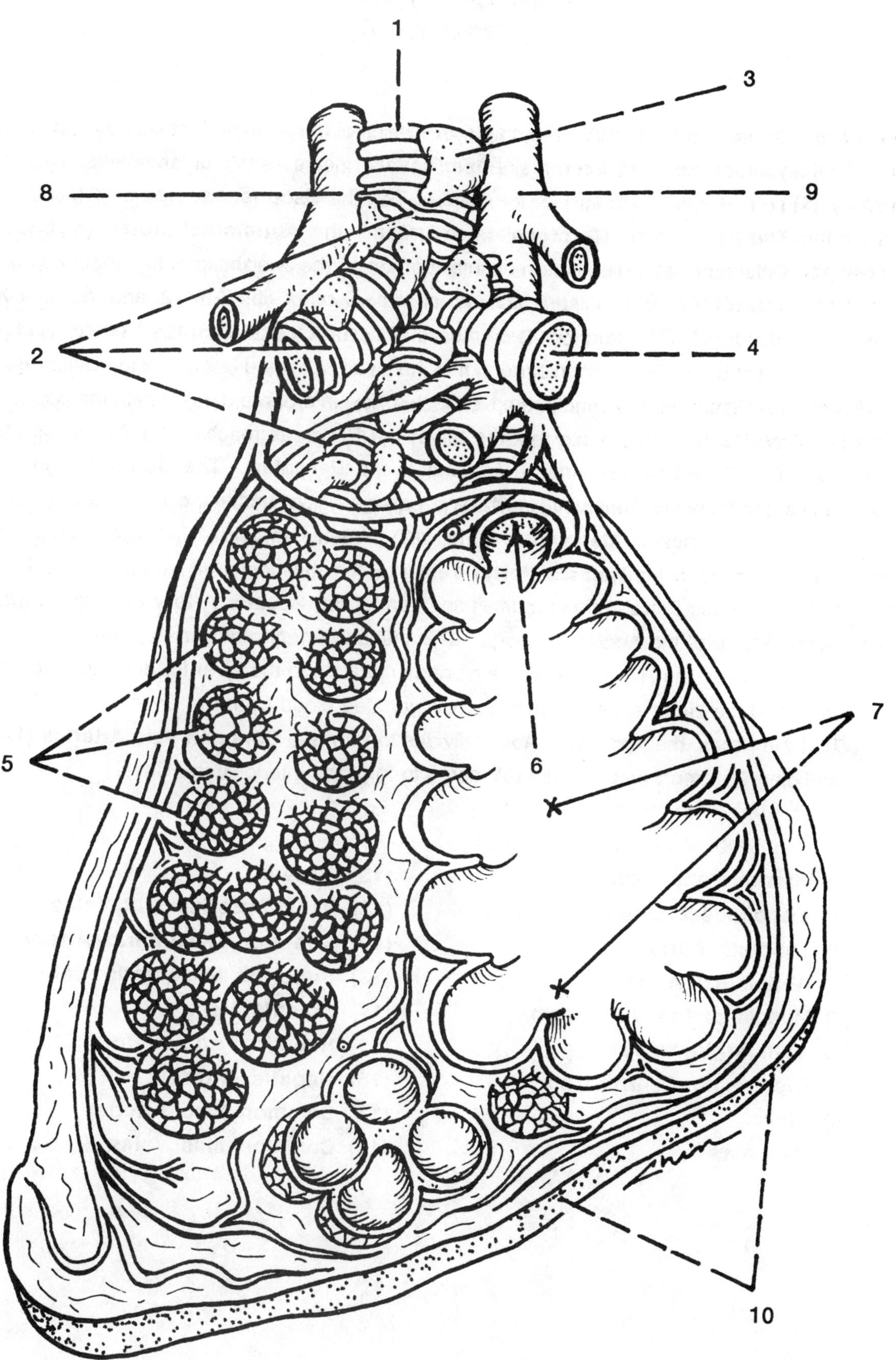

RS-4

Diaphragm
(Inferior View)

This view of the Diaphragm from below allows an appreciation of the abdominal structures that are located nearby. The Diaphragm contains three major openings that serve as useful landmarks. The **Inferior Vena Cava (1)** pierces the Diaphragm at the level of T-8, the **Esophageal Opening (2)** is opposite T-10, and the **Thoracic Aorta (3)** descends to become the **Abdominal Aorta (4)** once it has penetrated the Diaphragm at T-12. The muscular portion of the **Diaphragm (5)** originates from the **Lumbar Vertebrae (6)**, Costal Cartilages and adjacent portions of ribs 7 to 12, and from the Xiphoid Process of the Sternum. This great dome of muscular tissue inserts upon the **Central Tendon (7)** which is continuous with the Pericardium surrounding the Heart. Located just below the Diaphragm is the massive **Liver (8)** which has been largely removed in this view in order to demonstrate the position of the **Pancreas (9)** and **Spleen (10)**. Note that the tail of the Pancreas reaches the Spleen and that the **Renal Arteries and Veins (11)** emerge just below the Pancreas. The **Superior Mesenteric Artery (12)** branches off the Abdominal Aorta and emerges from under the Pancreas to supply the intestines, while the **Superior Mesenteric Vein (13)** joins the **Inferior Mesenteric Vein** and the **Splenic Vein (14)** to form the **Hepatic Portal Vein (15)** on the posterior surface of the Pancreas. To the left of the Hepatic Portal Vein is a small segment of the **Hepatic Artery (16)** which originated from the **Celiac Trunk** of the Abdominal Aorta. This Hepatic Artery provides oxygenated blood to the tissue of the Liver, while the Hepatic Portal Vein brings nutrients to the Liver for digestive processing to be completed, and the Hepatic Veins remove deoxygenated blood from the Liver to the Inferior Vena Cava. Within the Pelvis, the Abdominal Aorta divides to form the **Common Iliac Arteries (17)**, while the corresponding **Common Iliac Veins (18)** merge to form the Inferior Vena Cava.

(1) **Inferior Vena Cava**
(2) **Esophageal Opening**
(3) **Thoracic Aorta**
(4) **Abdominal Aorta**
(5) **Muscle of the Diaphragm**
(6) **Lumbar Vertebrae**
(7) **Diaphragm Central Tendon**
(8) **Liver**
(9) **Pancreas**
(10) **Spleen**
(11) **Renal Arteries and Veins**
(12) **Superior Mesenteric Artery**
(13) **Superior Mesenteric Vein**
(14) **Splenic Vein**
(15) **Hepatic Portal Vein**
(16) **Hepatic Artery**
(17) **Common Iliac Arteries**
(18) **Common Iliac Veins**

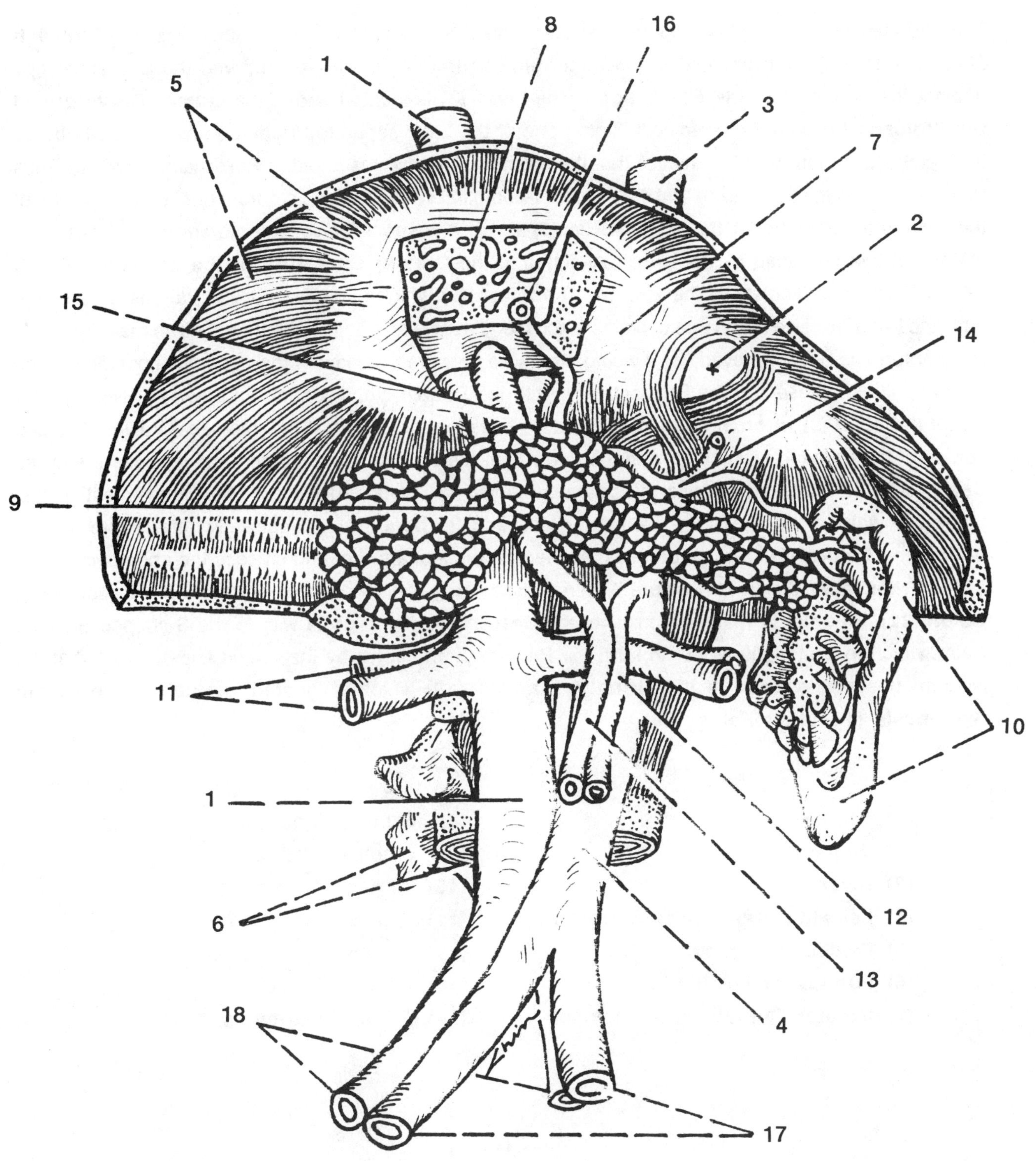

RS-5

Abdominal Cavity
(Superficial View)

The Abdominal Cavity has been opened for examination, with all of the vital organs intact. The **Stomach (1)** and **Spleen (2)** are situated in the upper left quadrant of the abdomen, with the Spleen resting against the curvature of the **Diaphragm**. The **Liver (3)** is located above the **Lesser Curvature of the Stomach (4)**, with the prominent Right Lobe of the Liver separated from the smaller Left Lobe by the **Falciform Ligament (5)**. Realize that the Falciform Ligament is a fold of Peritoneum which extends from the Liver to the Diaphragm and the Anterior Abdominal Wall. Peaking out from under the right lobe of the Liver is the **Fundus of the Gallbladder (6)** which is responsible for the storage and subsequent release of bile produced by the Liver. Attached to the **Greater Curvature of the Stomach (7)** is a fold of Peritoneum known as the **Greater Omentum (8)** which extends over the various coils of the Intestine like a massive apron that contains a rich blood supply and ample lymphatics. The majority of this large apron has been removed, for it would otherwise obscure our view of the abdominal contents. The food and liquids present in the Stomach pass into the Duodenum, Jejunum, and then the Ileum of the **Small Intestine (9)** which empties into the Large Intestine (or Colon) at the Ileo-cecal Valve. The first portion of the Large Intestine begins in the lower right quadrant of the abdomen and includes a sac-like Cecum with the attached Vermiform Appendix (shown by the dotted lines). The Large Intestine first passes upward along the right abdominal wall and is therefore appropriately referred to as the **Ascending Colon (10)**. The Large Intestine then makes a sharp turn and crosses the abdominal cavity as the **Transverse Colon (11)** before it then passes along the left abdominal wall as the **Descending Colon (12)**. In order to reach the midline, the Descending Colon gives way to the S-shaped Sigmoid Colon which then straightens out to form the Rectum (both shown by the dotted lines). Note that the Rectum passes posterior to the **Urinary Bladder (13)** which is located just behind the **Pubic Symphysis (14)** of the Pelvis.

(1) Stomach
(2) Spleen
(3) Liver
(4) Lesser Curvature of Stomach
(5) Falciform Ligament
(6) Fundus of Gallbladder
(7) Greater Curvature of Stomach
(8) Greater Omentum
(9) Small Intestine
(10) Ascending Colon
(11) Transverse Colon
(12) Descending Colon
(13) Urinary Bladder
(14) Pubic Symphysis

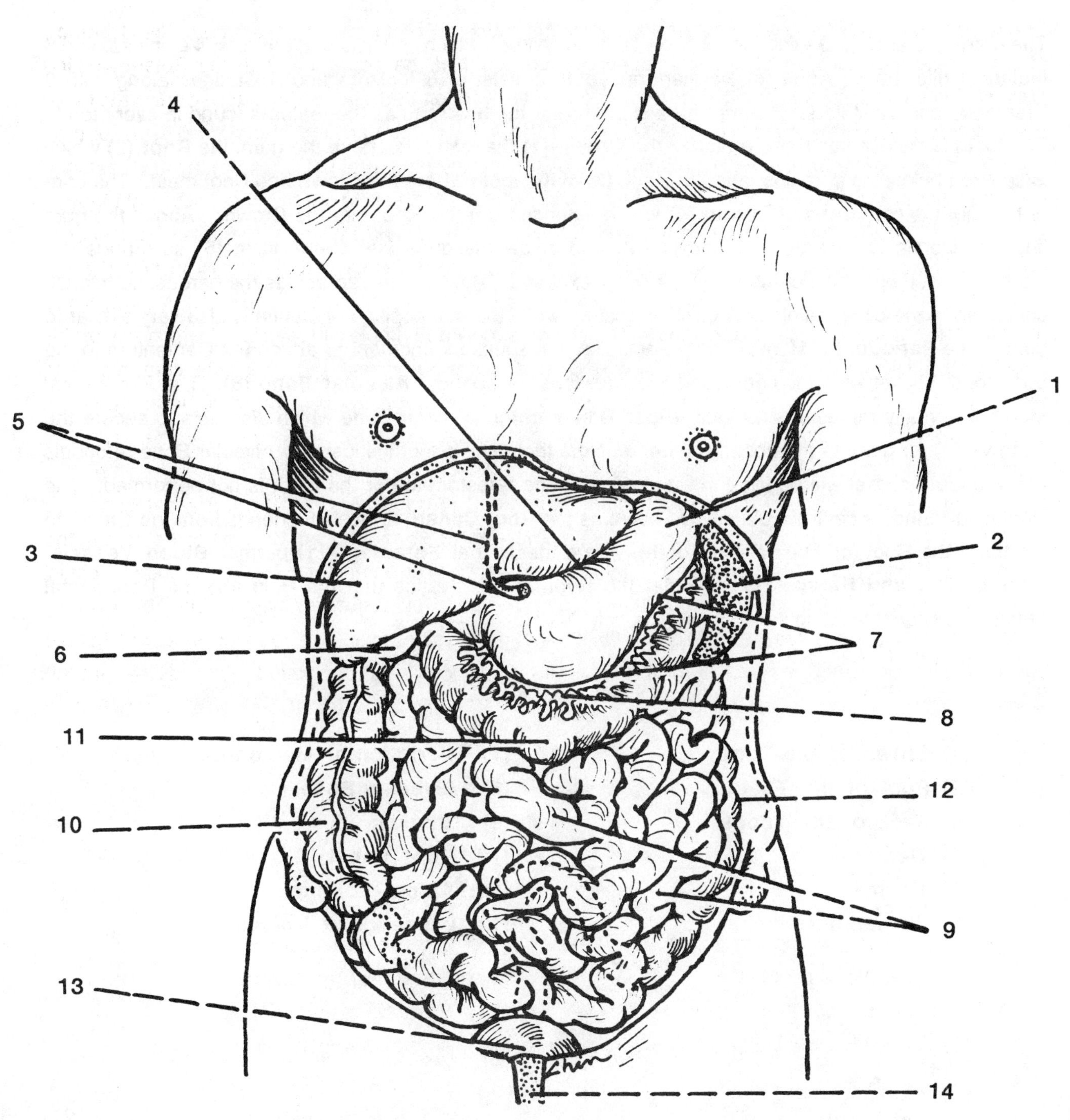

1
2
3
4
5
6
7
8
9
10
11
12
13
14

Tooth
Longitudinal Section

There are a total of 20 Deciduous (or Milk) Teeth in children which consists of 8 Incisors, 4 Canines, and 8 Molars, while the 32 Adult (or Permanent) Teeth consist of 8 Incisors and 4 Canines along with 8 Premolars and 12 Molars. Pictured here is an Incisor tooth which has the features found in every tooth. The tooth is divided into three regions: the **Crown (1)** that extends above the gum, the **Root (2)** which is secured below the gum line, and the **Neck (3)** of the tooth where the Crown and Root meet. The core of the tooth is composed of **Dentin (4)** which extends from the Root into the Crown. Above the gum line, the Dentin is covered by **Enamel (5)**, and below the gum line, **Cementum (6)** surrounds the tooth, so that the surface of the Dentin is never exposed. The Enamel represents the hardest substance contained in the body, being 97% calcified matter, while the composition of Dentin is also very similar to bone. The **Periodontal Membrane (7)** acts like a suspensory ligament to attach the Cementum to the bony tooth socket which is composed of cancellous (or spongy) **Alveolar Bone (8)**. The Periodontal Membrane really represents a modified periosteum of the Alveolar Bone which also acts to secure the **Gingiva (9)** (or gum) to the underlying bone. Note that once a tooth is lost, the Alveolar Bone remodels and the tooth socket will be filled in with spongy bone if restoration of the tooth is not performed. The Dentin surrounds a central Pulp cavity known as the **Root Canal (10)** which extends from the Crown to the tip of the Root (or Apex of the Tooth), where the **Apical Foramen (11)** permits **Blood Vessels, Lymphatics, and Nerves (12)** to enter the Root Canal. Notice that the Pulp and the Periodontal Membrane share the same sensory nerve supply.

(1) Crown of the Tooth
(2) Root of the Tooth
(3) Neck of the Tooth
(4) Dentin
(5) Enamel
(6) Cementum
(7) Periodontal Membrane
(8) Alveolar Bone
(9) Gingiva
(10) Root Canal
(11) Apical Foramen
(12) Vessels and Nerves

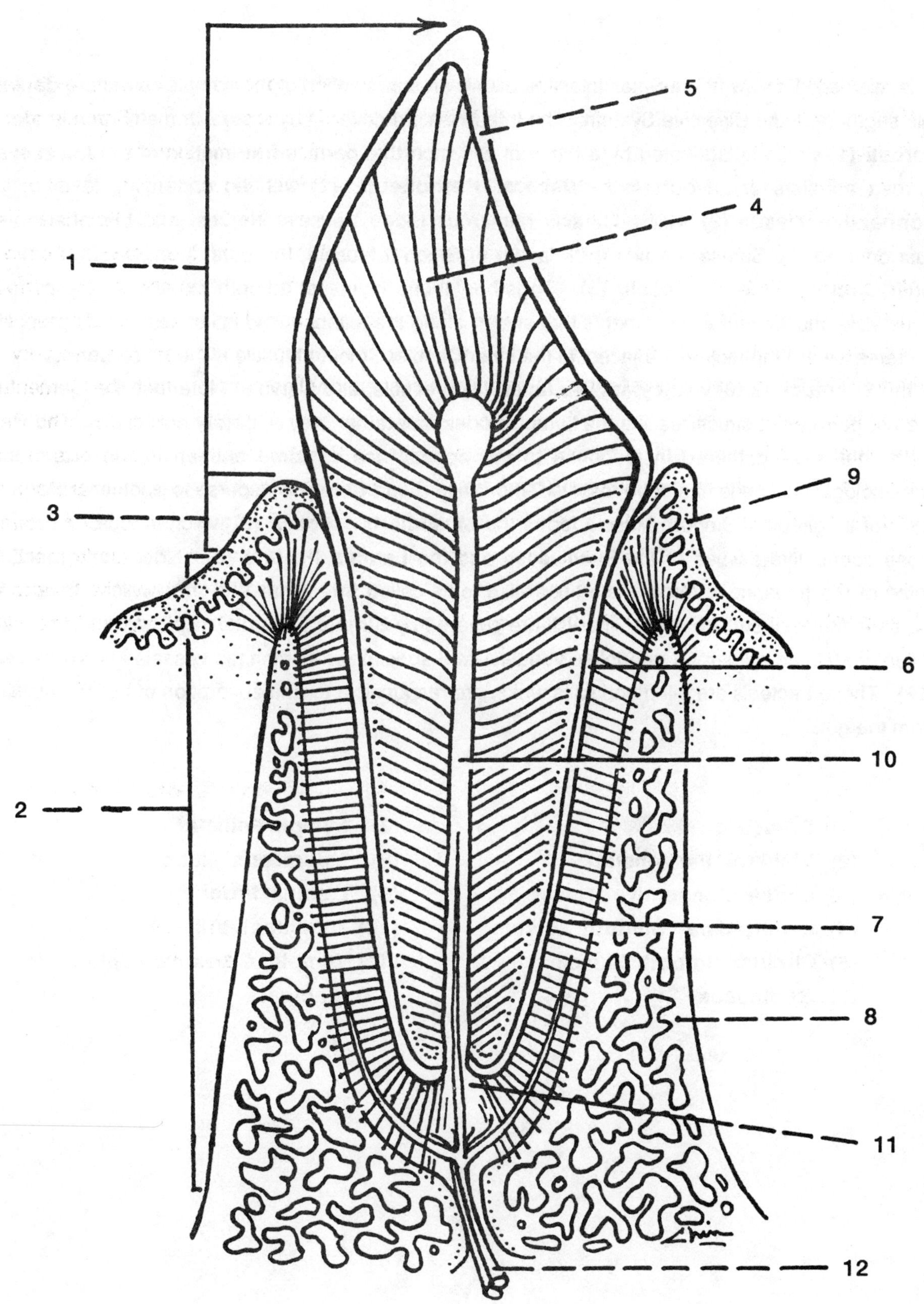

1
2
3
4
5
6
7
8
9
10
11
12

Histology of the Small Intestine

This microscopic view of the Small Intestine allows an appreciation of the complex structure displayed by this segment of the Digestive System. The Intestines are covered by a smooth membrane known as the **Serosa (1)** which is lubricated by a thin film of lymph that permits free motion of the Intestines. The Serosa consists of an outermost **Visceral Peritoneum (2)** with an underlying layer of **Loose Connective Tissue (3)** which contains numerous Blood Vessels, Nerves, and Lymphatic Vessels. Just deep to the Serosa are two thick layers of Smooth Muscle, the outer **Longitudinal (4)** and the inner **Circular Smooth Muscle (5).** Together, these layers of smooth muscle are responsible for Peristalsis, the wave-like movements that sweep along the extent of the intestines, slowly propelling the contents toward the rectum. Adjacent to the inner Circular Smooth Muscle is the loose Connective Tissue of the **Submucosa (6)** with **Lymph Follicles (7)**, Vessels, and Nerves. Note that the Lymph Follicles are not permanent structures like the Lymph Nodes, but rather are temporary collections of White Blood Cells that have gathered in response to the appearance of some antigen in the gut. Once the immunologic challenge has been resolved, the White Blood Cells will disperse to another location. A thin layer of longitudinal Smooth Muscle forms the **Muscularis Mucosa (8)** which is located between the loose connective tissues of the Submucosa and the **Lamina Propria (9)**. Notice that the Epithelial lining of the Intestine is thrown into finger-like folds called **Intestinal Villi (10)** which serve to greatly increase the total surface area. The Villi are covered by **Simple Columnar Epithelium (11)**, with each Villus containing numerous arterioles, venules, and specialized Lymphatic vessels known as **Lacteals (12)**. These Lacteals are blind-end sacs which are responsible for the absorption of bulky emulsified fats from the gut.

(1) Serosa
(2) Visceral Peritoneum
(3) Loose Connective Tissue
(4) Longitudinal Smooth Muscle
(5) Circular Smooth Muscle
(6) Submucosa
(7) Lymph Follicle
(8) Muscularis Mucosa
(9) Lamina Propria
(10) Intestinal Villi
(11) Simple Columnar Epithelium
(12) Lacteal

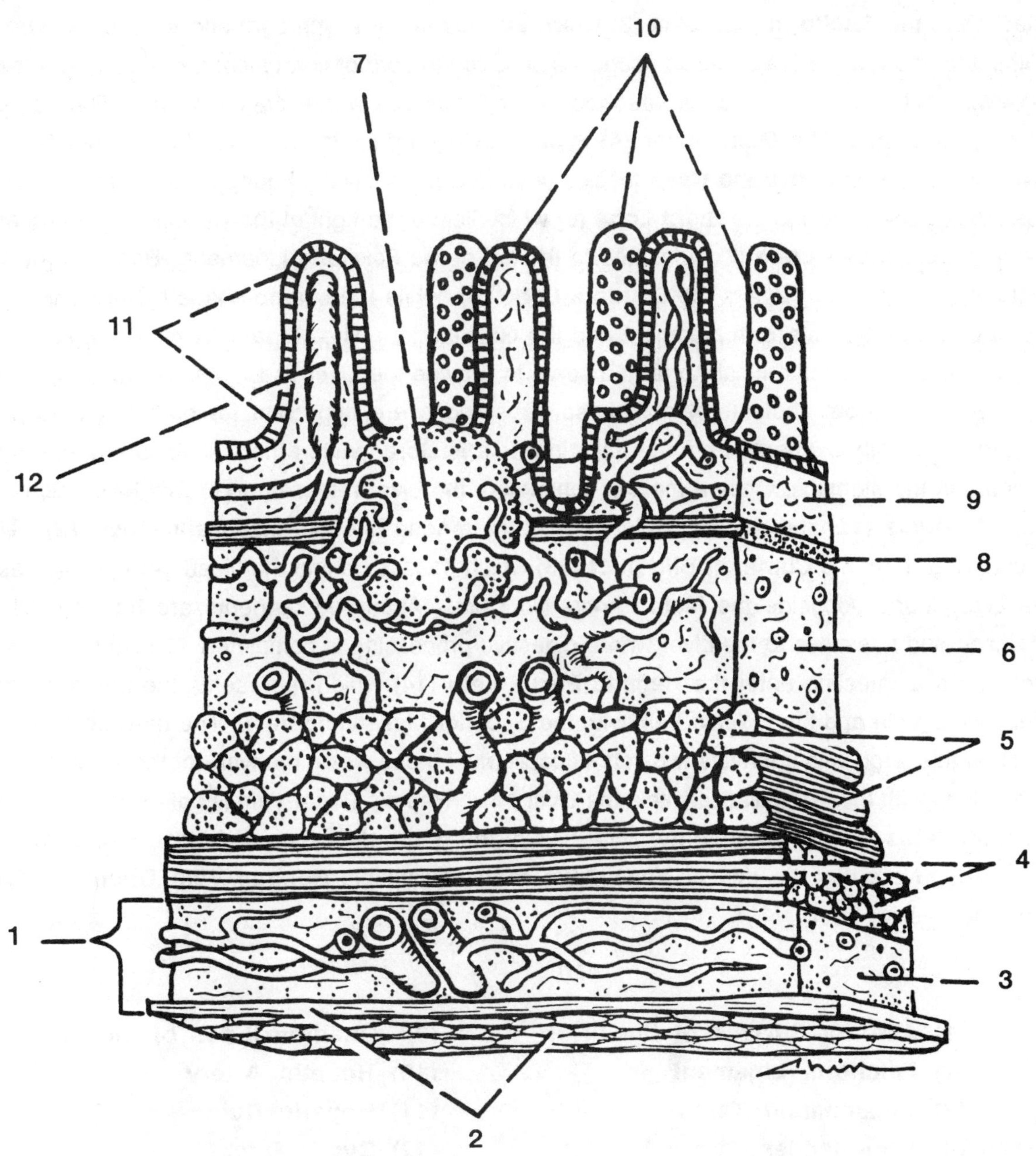
10
7
11
12
9
8
6
5
4
1
3
2

Liver
(Inferior View)

The Liver is the largest gland within the body, serving as an important component of the Digestive System because of its role in the production of bile and the subsequent metabolism of absorbed nutrients from the gut. The undersurface of the Liver is evident here, with a portion of the soft tissues removed in order to display the associated vessels and ducts. The dome-shaped **Anterior Surface of the Liver (1)** is attached to the **Falciform Ligament (2)** which extends to the Diaphragm and Anterior Abdominal Wall. Within the free lower edge of the ligament is a solid round cord of fibrous connective tissue known as the **Ligamentum Teres (3)** which is attached to the Umbilicus and represents the obliterated Umbilical Vein of the Fetus. The **Gallbladder (4)** is seen extending slightly beyond the Anterior Border of the Liver, while the **Inferior Vena Cava (5)** lies posterior to the Liver. Four Lobes of the Liver have been described. The largest is the **Right Lobe (6)** which lies to the right of the Gallbladder and Inferior Vena Cava, while the **Left Lobe (7)** is located to the left of the Falciform Ligament. Between the Left Lobe and the Inferior Vena Cava is the **Caudate Lobe (8)**, while the square-shaped **Quadrate Lobe (9)** is found between the Falciform Ligament and the Gallbladder. The **Hepatic Artery (10)** arises from the Celiac Trunk of the Abdominal Aorta and serve to provide the Liver tissue with oxygen. The Liver cells form bile salts from cholesterol and these substances are important for both the emulsifying and absorption of fatty elements from the intestine. Bile is stored within the Gallbladder until food actually appears in the stomach. Once a meal is consumed, the Liver releases additional Bile which enters the **Hepatic Ducts (11)** and the Gallbladder contracts to send Bile into the **Cystic Duct (12)**. These two ducts join to form the **Common Bile Duct (13)** which passes behind the Head of the Pancreas to enter the Duodenum alongside the Main Pancreatic Duct. Absorbed nutrients are transported from the Stomach and Intestines to the Liver where digestive processing is completed. These nutrients from the stomach and intestine enter the **Hepatic Portal Vein (14)** which represents the union of the Inferior Mesenteric Vein and the Splenic Vein with the Superior Mesenteric Vein. The term *portal* refers to the fact that this large vein connects two sets of capillaries: those within the walls of the stomach, intestines, and spleen with those capillaries within the Liver. The Hepatic Portal Vein is not to be confused with the **Hepatic Veins (15)** which drain the Liver tissue and return deoxygenated blood to the **Inferior Vena Cava (5)**. Note that the Inferior Vena Cava contains multiple **Hepatic Vein Openings (16)** which drain the massive Liver.

(1) Anterior Surface of the Liver
(2) Falciform Ligament
(3) Ligamentum Teres
(4) Gall-bladder
(5) Inferior Vena Cava
(6) Right Lobe of the Liver
(7) Left Lobe of the Liver
(8) Caudate Lobe of the Liver
(9) Quadrate Lobe of the Liver
(10) Hepatic Artery
(11) Hepatic Duct
(12) Cystic Duct
(13) Common Bile Duct
(14) Hepatic Portal Vein
(15) Hepatic Vein
(16) Hepatic Vein Openings

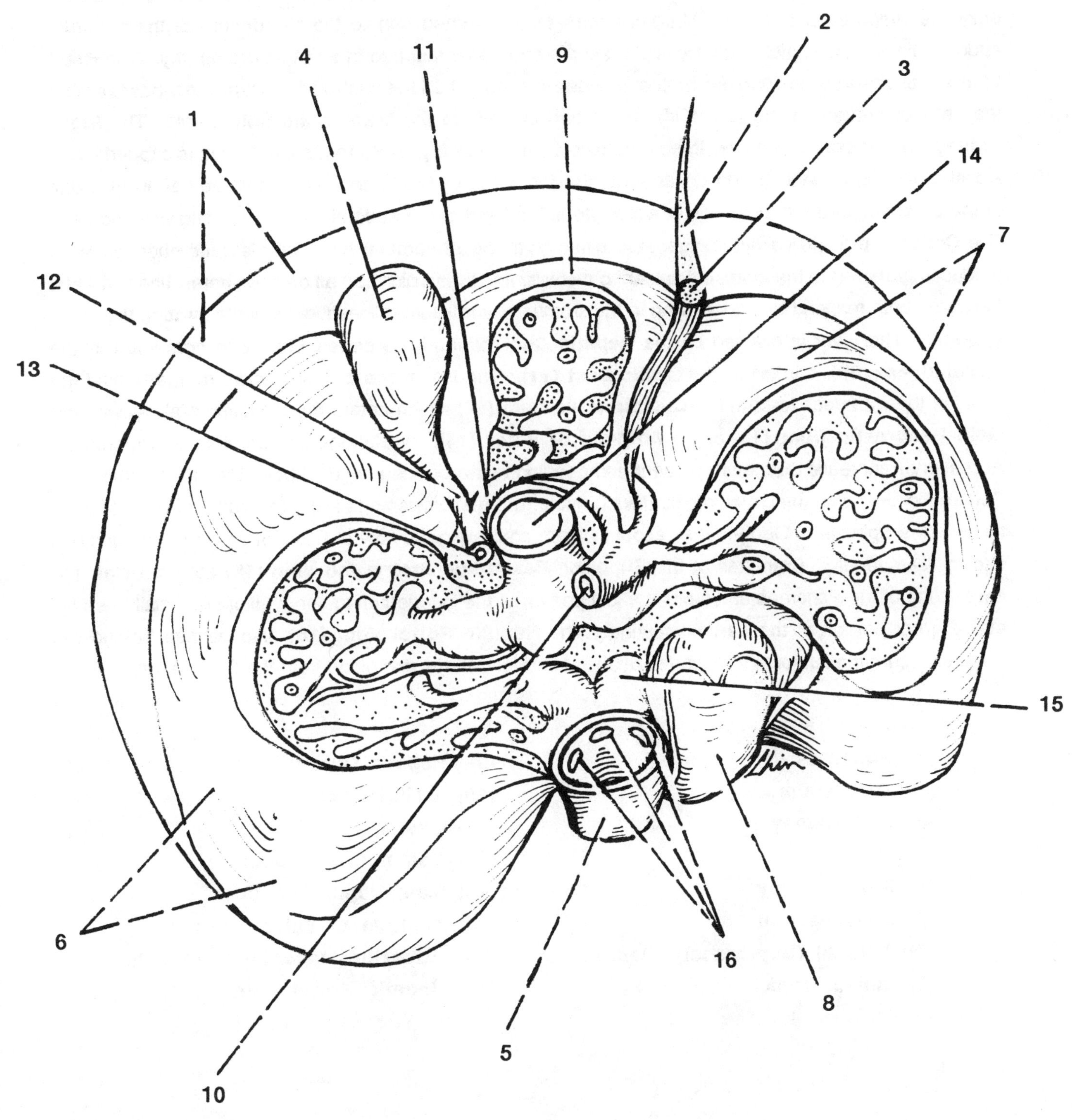

DS-4

Relations of the Pancreas

The Pancreas serves as both an Endocrine Organ (producing various blood-born elements such as Insulin and Glucagon) and an Exocrine Organ (responsible for the production and release of digestive enzymes which enter the gut). The **Pancreas (1)** is situated behind the peritoneum of the posterior abdominal wall and is pictured here to demonstrate the orientation to other structure nearby. The **Head of the Pancreas** is surrounded by the **Duodenum (2)**, while the **Tail of the Pancreas** passes over the anterior surface of the **Left Kidney (3)** and extends to the hilum of the **Spleen (4)**. The **Right Kidney (5)** is in contact with the inferior surface of the **Liver (6)**. Note that each Kidney is capped by an **Adrenal** (or Suprarenal) **Gland (7)** and that the Right Kidney is situated lower than the Left Kidney due to the presence of the massive Liver which stops it ascent in the abdominal cavity during development. The **Celiac Trunk (8)** carries oxygenated blood from the Abdominal Aorta and is quite short before it produces branches to the stomach, spleen, pancreas, and liver which can all be seen here. The Liver and Pancreas both respond to food arriving in the stomach by releasing digestive elements through their duct systems. The Liver is drained by the **Hepatic Duct (9)** and the stored Liver Bile contained in the **Gallbladder (10)** is carried by the **Cystic Duct (11)**. The Hepatic and Cystic Ducts merge behind the Head of the Pancreas to form the **Common Bile Duct (12)** which enters the Duodenum along with the **Main Pancreatic Duct (13)** at the **Ampulla of Vater (14)**. Variations are common here, with an Accessory Pancreatic Duct often present and opening in the Duodenum just above the union of the Main Pancreatic Duct with the Common Bile Duct. Note that these openings are guarded by circular smooth muscle (the Spincter of Oddi) that is responsible for controlling the flow of Bile and Pancreatic secretions into the Duodenum. Branches of the **Superior Mesenteric Artery and Vein (15)** appear below the lower edge of the Pancreas and cross the last portion of the Duodenum to serve all of the Small Intestine and a great portion of the Large Intestine. The **Hepatic Portal Vein (16)** also passes behind the Pancreas, carrying nutrients transported by the Mesenteric Veins and blood from the Splenic Vein.

(1) Pancreas
(2) Duodenum
(3) Left Kidney
(4) Spleen
(5) Right Kidney
(6) Liver section
(7) Adrenal (Suprarenal) Glands
(8) Celiac Trunk

(9) Hepatic Duct
(10) Gallbladder
(11) Cystic Duct
(12) Common Bile Duct
(13) Main Pancreatic Duct
(14) Ampulla of Vater
(15) Superior Mesenteric Vessels
(16) Hepatic Portal Vein

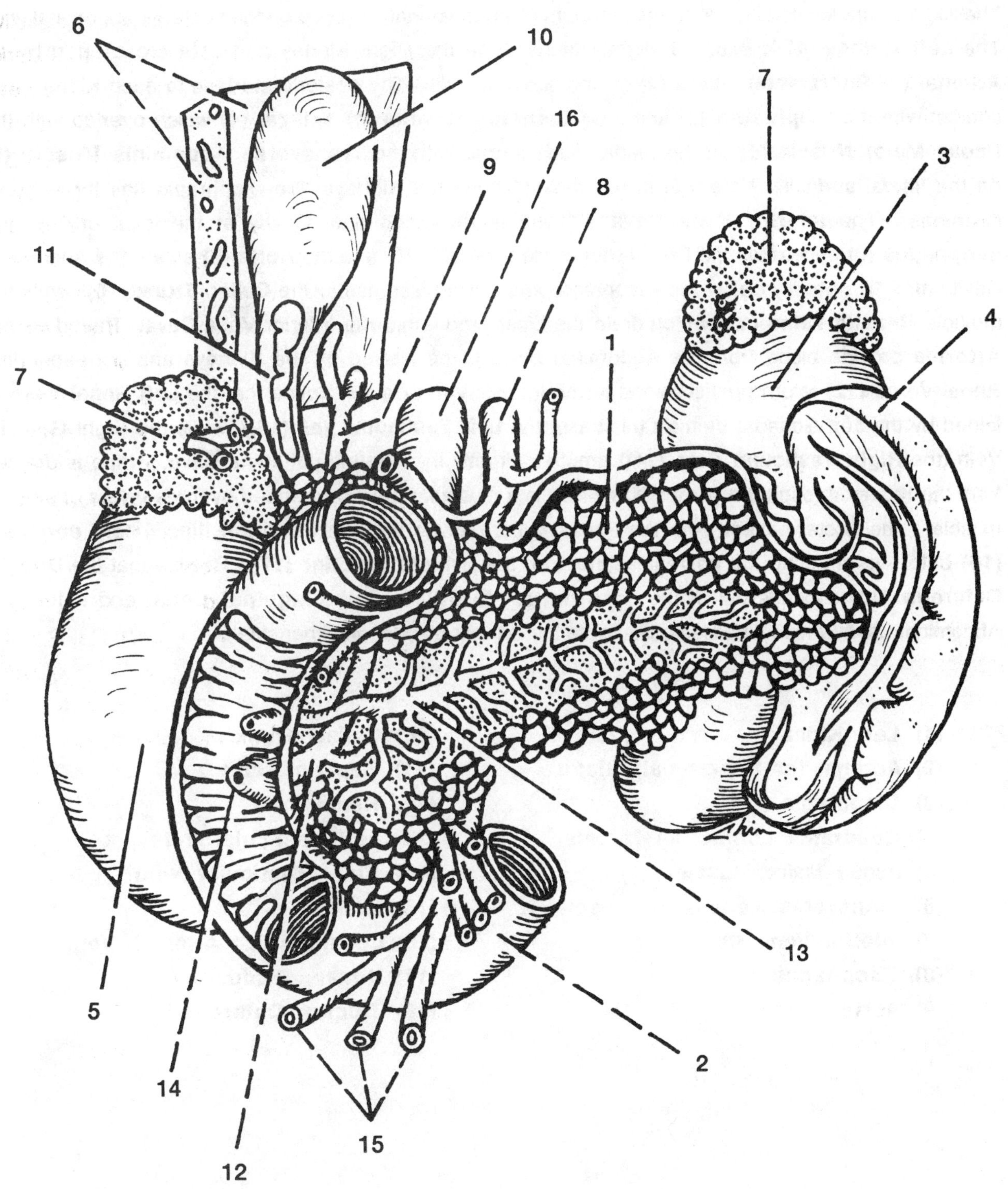

DS-5

Urinary Tract Elements

(Male)

The Kidneys are located behind the peritoneum (retroperitoneal), high up on the posterior abdominal wall. The **Left Kidney (1)** is situated slightly higher than the Right Kidney and both are capped by an **Adrenal (or Suprarenal) Gland (2)** on the superior pole. The posterior surface to each Kidney is in contact with the **Diaphragm (3)** and the **Quadratus Lumborum Muscle (4)** which overlap with the **Psoas Major Muscle (5)** on the medial surface and with the **Transverse Abdominis Muscle (6)** on the lateral surface. Note that in the region above the Kidneys, the Diaphragm has three major openings. The **Inferior Vena Cava (7)** leaves the Abdominal Cavity at the level of T-8, the **Esophagus (8)** penetrates the Diaphragm at the level of T-10, and the **Aorta (9)** enters the Abdominal Cavity at T-12. Shortly below the Diaphragm, the Aorta gives rise to the **Celiac Trunk (10)**, while the multiple **Hepatic Veins (11)** which drain the Liver empty into the Inferior Vena Cava. Paired **Renal Arteries** provide blood from the Abdominal Aorta to be filtered by the Kidneys and corresponding **Renal Veins (12)** return purified blood to the Inferior Vena Cava. Notice that the **Left Renal Vein** is joined by the Left Gonadal Vein (in this case, the **Left Testicular Vein(13)**), while the Right Gonadal Vein (the **Right Testicular Vein (14)**) empties directly into the Inferior Vena Cava. Urine is drained from the Kidney through the **Ureter (15)** which is a muscular tube surrounded by three layers of smooth muscle. The Ureter passes over the Psoas Major muscle and the **Common Iliac Artery and Vein (16)** before penetrating the posterior surface of the **Urinary Bladder (17)**. Realize that the **Ductus Deferens (18)** ascends out of the **Scrotum** to travel through the **Inguinal Canal** and enters the Abdominal Cavity where it passes posterior to the Urinary Bladder on either side.

(1) Left Kidney
(2) Adrenal (or Suprarenal) Gland
(3) Diaphragm
(4) Quadratus Lumborum Muscle
(5) Psoas Major Muscle
(6) Transverse Abdominis Muscle
(7) Inferior Vena Cava
(8) Esophagus
(9) Aorta
(10) Celiac Trunk
(11) Hepatic Veins
(12) Renal Vein
(13) Left Testicular Vein
(14) Right Testicular Vein
(15) Ureter
(16) Common Iliac Artery & Vein
(17) Urinary Bladder
(18) Ductus Deferens

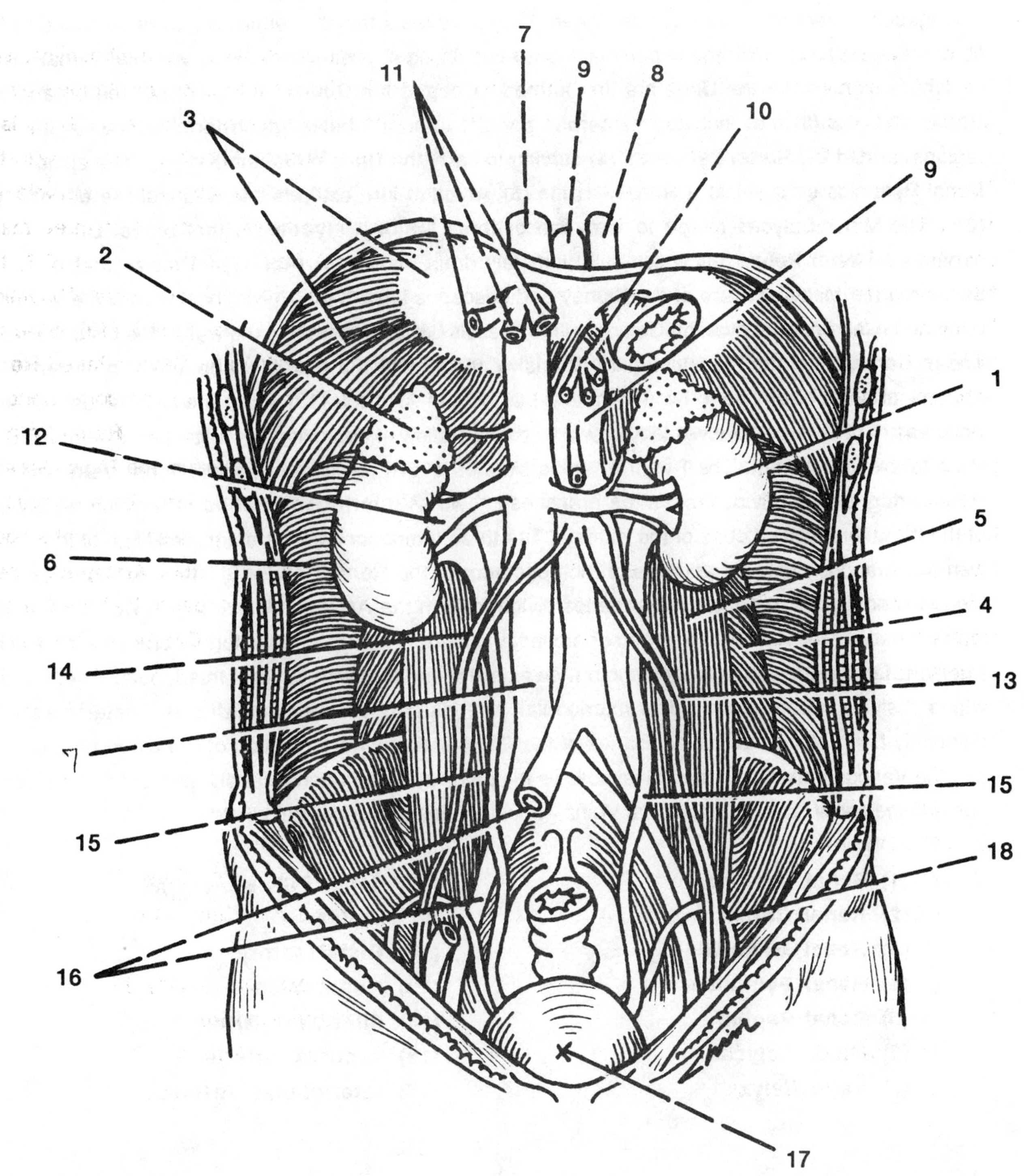
7
11
9
8
10
3
9
2
1
12
5
6
4
14
13
7
15
15
18
16
17

Kidney

(Interior View with Arterial Blood Supply)

A longitudinal section of the retroperitoneal Kidney reveals that the entire organ is surrounded by a fibrous **Capsule (1)**, with the superficial kidney consisting of a dark red, highly vascular **Renal Cortex (2)** where elements of the Urine are first formed. Deep to the Cortex, in the core of the organ, is the **Renal Medulla (3)** that includes numerous parallel urinary tubules which form somewhat triangular regions termed the **Renal Pyramids (4)** serving to carry the Urine out of the Kidney. The apices of the Renal Pyramids empty into a **Renal Papilla (5)**, which in turn extends into a funnel-like **Minor Calyx (6)**. The Minor Calyces merge to form two or three **Major Calyces (7)** that propel Urine into the expanded **Renal Pelvis (8)** which will ultimately drain into the **Ureter (9)**. Realize that the Ureter begins on the Medial surface of the Kidney and descends toward the Pelvis, posterior to the abdominal peritoneum, and passes over the Common Iliac Vessels before penetrating the wall of the Urinary Bladder. The blood supply of the Kidney is quite extensive and it follows a consistent pattern of branching. The Abdominal Aorta gives rise to two (Right and Left) **Renal Arteries (10)** which are located just below the Pancreas. These large-bore vessels allow a Renal blood flow of approximately 1 liter per minute (or 20% of the total blood volume). The Renal Artery is situated above the corresponding **Renal Vein (11)**. The Renal Artery usually divides into three branches known as **Interlobar Arteries (12)** which supply each of the three embryonic Lobes of the Kidney. The three Interlobar Arteries enter the Hilum of the Kidney, with two branches anterior and one branch posterior to the **Renal Pelvis (8)**. Near the junction of the Medulla and Cortex, the Interlobar Arteries divide to form the **Arcuate Arteries (13)** which run parallel to the surface of the Kidney in a bowed or arched array. The Arcuate Arteries further branch it form smaller **Interlobular Arteries (14)** found running towards the surface, with several Interlobular Arteries found within a single Kidney Lobe. The Interlobular Arteries connect with the **Afferent Arterioles** which ultimately bring blood into the capillaries of the **Glomerulus** where filtration of the blood begins. Note that the Venous drainage of the Kidney follows the pathway of the arterial supply, with each of the vessels carrying the same names (Interlobular Veins, Arcuate Veins, Interlobar Veins, etc).

(1) Renal Capsule
(2) Renal Cortex
(3) Renal Medulla
(4) Renal Pyramids
(5) Renal Papilla
(6) Minor Calyces
(7) Major Calyx
(8) Renal Pelvis
(9) Ureter
(10) Renal Artery
(11) Renal Vein
(12) Interlobar Artery
(13) Arcuate Arteries
(14) Interlobular Arteries

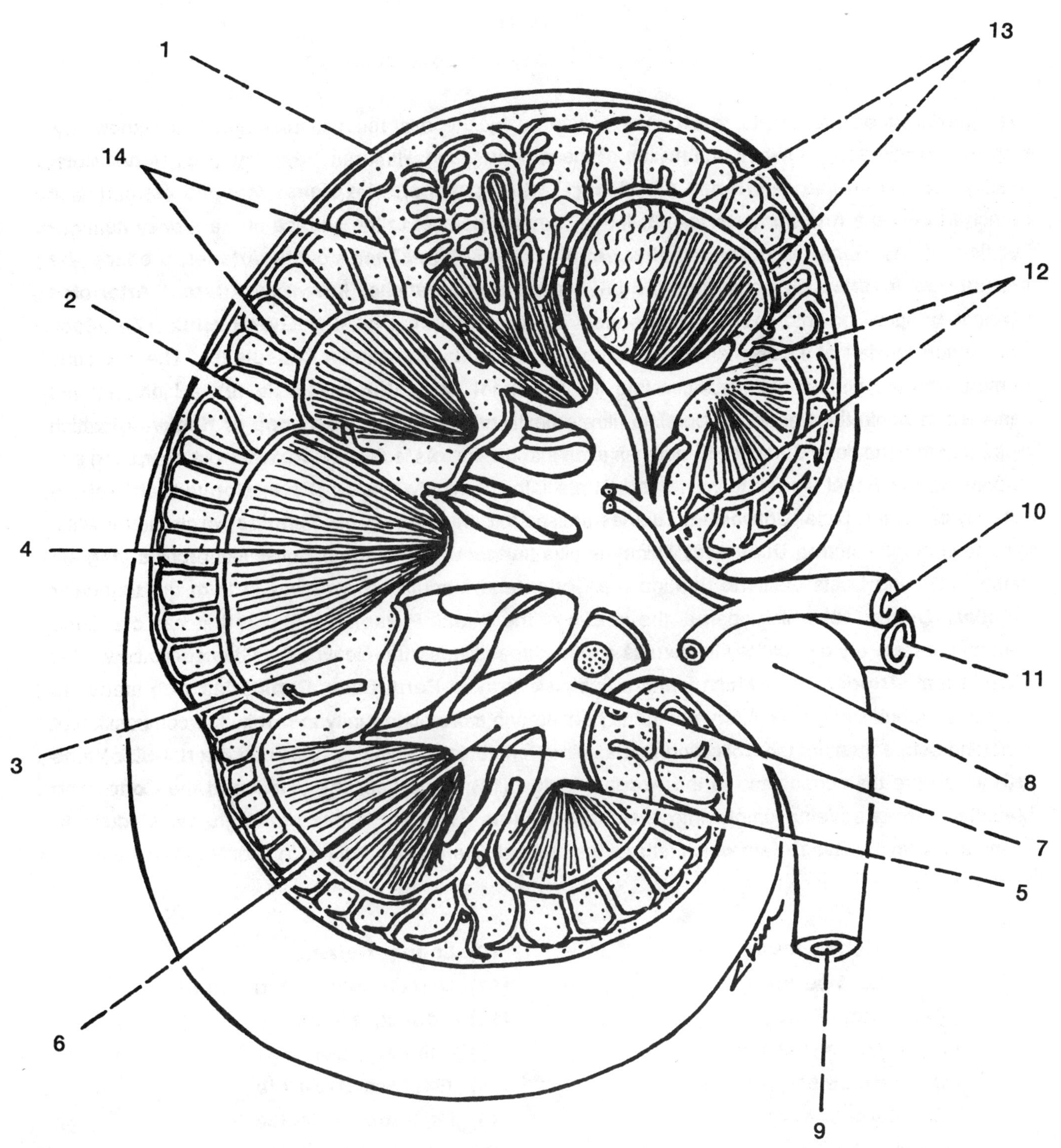
1
13
14
12
2
10
4
11
8
3
7
5
6
9

The Nephron

The arterial blood brought to the Kidney is filtered by the microscopic functional units known as **Nephrons**, with approximately one million of these units found within each Kidney. The Renal Artery usually produces three **Interlobar Arteries (1)** which penetrate the **Renal Medulla (2)** and then branch to form the **Arcuate Arteries (3)** found running parallel to the surface of the Kidney near the junction of the Renal Medulla and the **Renal Cortex (4)**. The Arcuate Arteries produce the **Interlobular Arteries (5)** which then display multiple small branches known as **Afferent Arterioles (6)** that bring blood into the first component of the Nephron Unit, the **Glomerulus (7)**. Each Glomerulus consists of a collection of capillaries surrounded by **Bowman's Capsule (8)**. The first Urine formed from the filtered blood passes into the **Proximal Convoluted Tubule (9)**, a highly curved canal found within the Renal Cortex. The filtrate next enters the U-shaped **Loop of Henle (10)** which dips down into the deep Cortex or sometimes even into the Renal Medulla before making a sharp turn and re-entering the Renal Cortex to next send filtrate into the highly twisted **Distal Convoluted Tubule (11)**. Urine in the Distal Convoluted Tubules passes into the **Collecting Tubules** which in turn drain into the larger **Collecting Ducts (12)** which receive a vast number of other Collecting Tubules. A great many Collecting Ducts descend through the Cortex and Medulla, converging to form the triangular **Papillary Ducts (13)** which open at the apices of the **Renal Pyramids**. Blood that has circulated through the Capillary bed of the Glomerulus and has been filtered then leaves the Glomerulus by way of the **Efferent Arteriole (14)** which forms a complex array of **Peritubular Capillaries (15)** that wrap around the tubules of the Nephron and act to provide one more opportunity to modify the composition of the Urine before it enters the Collecting Ducts. The Peritubular Capillaries drain into **Interlobular Veins (16)** which are then connected to the **Arcuate Veins (17)** found near the junction of the Cortex and Medulla. Arcuate Veins empty into **Interlobar Veins (18)** which pass through the Medulla to eventually form the **Renal Vein** which returns the purified, filtered blood to the Inferior Vena Cava.

(1) Interlobar Artery
(2) Renal Medulla
(3) Arcuate Artery
(4) Renal Cortex
(5) Interlobular Artery
(6) Afferent Arteriole
(7) Glomerulus
(8) Bowman's Capsule
(9) Proximal Convoluted Tubule
(10) Loop of Henle
(11) Distal Convoluted Tubule
(12) Collecting Duct
(13) Papillary Duct
(14) Efferent Arteriole
(15) Peritubilar Capillaries
(16) Interlobular Vein
(17) Arcuate Vein
(18) Interlobar Vein

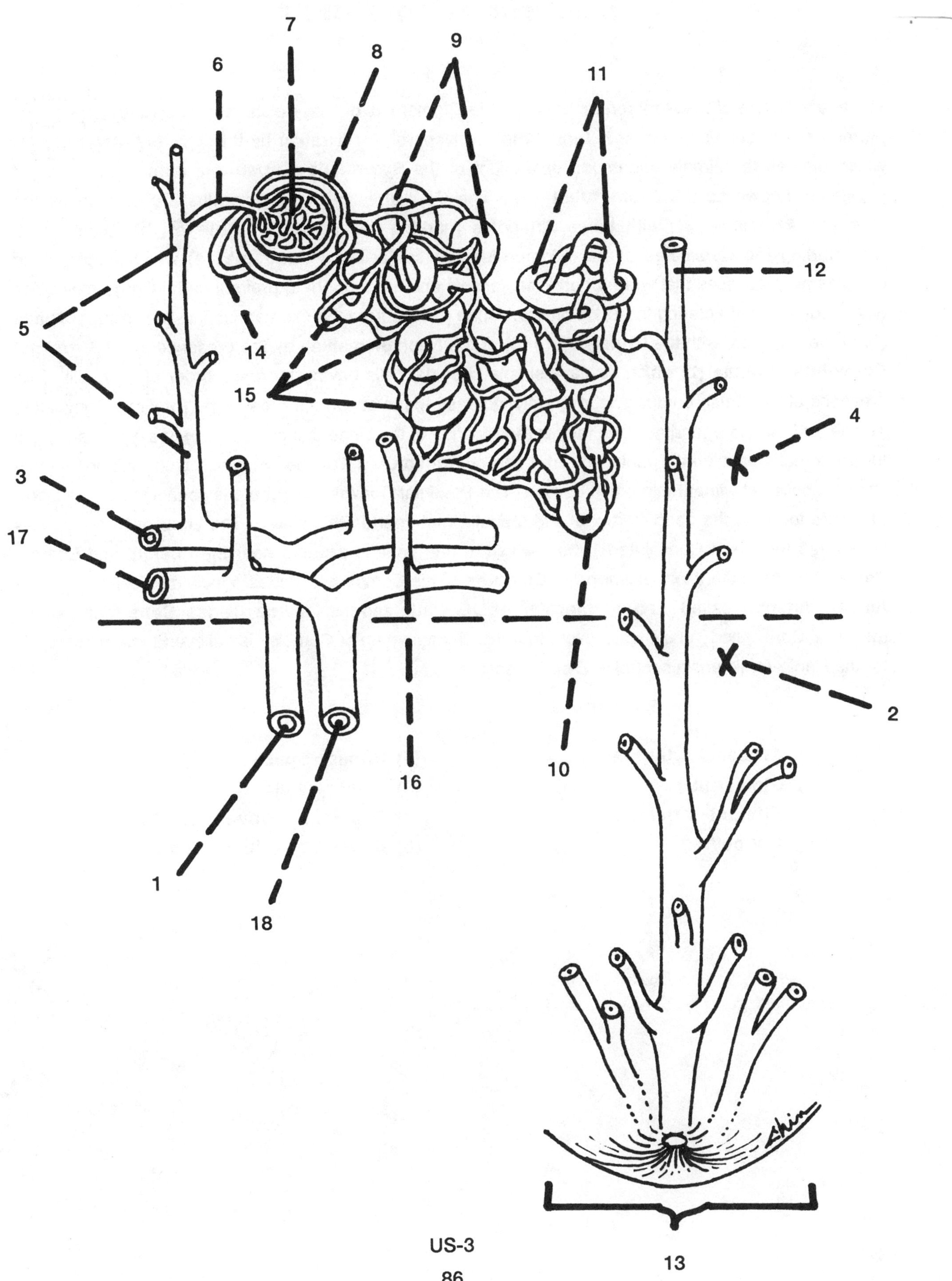

US-3

Histology of the Nephron Glomerulus

The fine structure of the Nephron reveals that the Glomerulus displays distinct histologic features which permit easy orientation. Blood is brought to the Nephron for filtration by the **Afferent Arteriole (1)** which pierces the simple squamous epithelium of the **Bowman's Capsule** to form a collection of capillaries known as the **Glomerulus (2)**. The Capillaries of the Glomerulus are drained by the **Efferent Arteriole (3)**, with these Arterioles marking the **Vascular Pole** of the Glomerulus. Surrounding the Capillaries of the Glomerulus are spider-like **Podocytes (4)** which display long cytoplasmic processes that wrap around the vessels and are thought to prevent certain large molecules (like Albumin) from entering the Urine filtrate. Urine passes from the Glomerulus into the **Urinary Space (5)** of the Nephron, with the **Urinary Pole (6)** of the Nephron marked by the presence of the **Proximal Convoluted Tubule (7)** which displays simple cuboidal cells having a brush border of microvilli. The presence of this brush border acts to greatly increase the total surface area of the Proximal Convoluted Tubule and therefore facilitates the further modification of the Urine composition. It is easy to distinguish the Proximal Convoluted Tubule from the **Distal Convoluted Tubule (8)** which also includes simple cuboidal cells yet without this brush border. The Proximal Convoluted Tubule is connected to the **Loop of Henle** found in the deep Cortex or the Medulla (not seen in this view). The Loop of Henle passes filtrate into the Distal Convoluted Tubule which at one point is situated near the Afferent and Efferent Arterioles to form the **Juxtaglomerular Complex**. This Complex includes specialized cells within the Afferent Arteriole called **Juxtaglomerular or JG Cells** and the **Macula Densa Cells** found within the Distal Convoluted Tubule wall. Note that this Juxtaglomerular Complex is believed to be responsible for the Renin-Angiotensin control of Blood Pressure.

(1) Afferent Arteriole
(2) Glomerulus
(3) Efferent Arteriole
(4) Podocytes
(5) Urinary Space
(6) Urinary Pole
(7) Proximal Convoluted Tubule
(8) Distal Convoluted Tubule

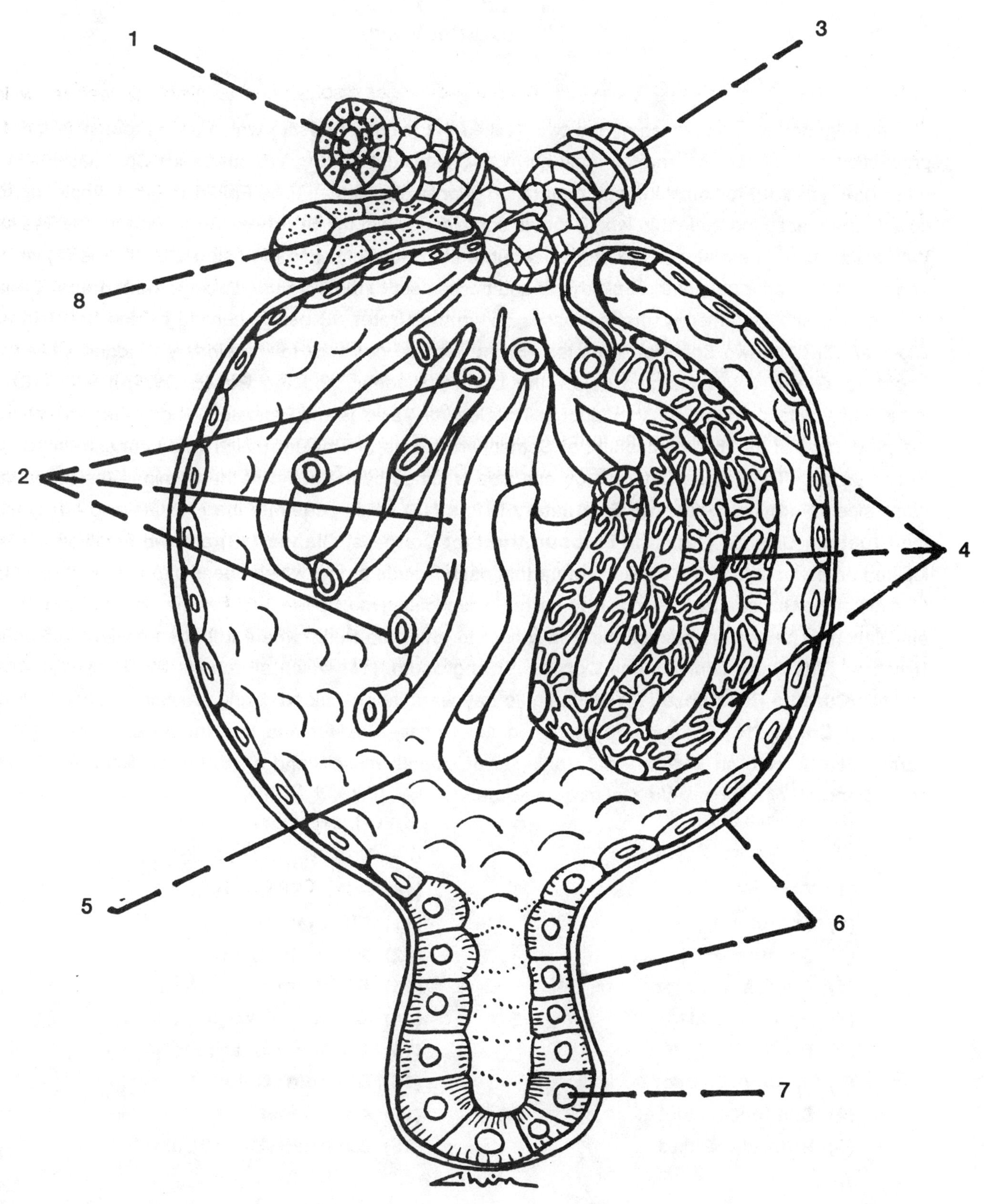

US-4

Male Pelvis
(Midsagittal View)

The male gonad is represented by the two **Testes (1)** (singular, testis) which are responsible for the production of Spermatozoa and the male hormone Testosterone. Each Testis sends Spermatozoa into one **Epididymis (2)** to complete their maturation prior to ejaculation. The Epididymis is a single, highly coiled tube measuring 20 feet in length that is contained within the **Scrotum (3)**, along with the Testes. With ejaculation, Spermatozoa are propelled out of the Epididymis into the **Ductus Deferens (4)** which begins at the inferior pole of the Epididymis and passes out of the Scrotum, through the **Inguinal Canal (5)**, and into the Abdomen where is crosses above the **Ureter (6)** before curving behind the **Urinary Bladder (7)**. The two **Seminal Vesicles (8)** are located posterior to the Urinary Bladder, while the **Prostate Gland (9)** is situated below the Urinary Bladder, and the **Pubic Symphysis (10)** is immediately anterior to the Urinary Bladder. The Seminal Vesicles and Prostate Gland are responsible for the production of the fluid component of Semen which acts to provide motility and nourishment to the Spermatozoa. The Ductus Deferens on one side is joined by the outlet of the Seminal Vesicle on the same side to form one of the two **Ejaculatory Ducts (11)** which penetrate the Prostate and empty into the **Prostatic Urethra (12)**. The **Bulbourethral** (or Cowper's) **Glands (13)** are two small structures located on either side of the Penile Urethra that each include a duct which opens into the Urethra. The function of the Bulbourethral Glands is to secrete a lubricating mucus-like fluid into the Urethra just prior to ejaculation. The Urethra pierces the Pelvic floor to enter the Penis where it is then called the **Penile Urethra**. The Penis contains one **Corpus Spongiosum (14)** which surrounds the Penile Urethra, and two **Corpora Cavernosa (15)** which act to trap warm arterial blood during erection. Notice that the **Sigmoid Colon (16)** is located posterior to the Urinary Bladder and that the **Anal Canal (17)** is surrounded by both an Internal Anal Sphincter of smooth muscle and an **External Anal Sphincter (18)** composed of voluntary skeletal muscle tissue.

(1) Testis
(2) Epididymis
(3) Scrotum
(4) Ductus Deferens
(5) Inguinal Canal
(6) Ureter
(7) Urinary Bladder
(8) Seminal Vesicle
(9) Prostate Gland
(10) Pubic Symphysis
(11) Ejaculatory Duct
(12) Prostatic Urethra
(13) Bulbourethral Glands
(14) Corpus Spongiosum
(15) Corpus Cavernosa
(16) Sigmoid Colon
(17) Anal Canal
(18) External Anal Sphincter

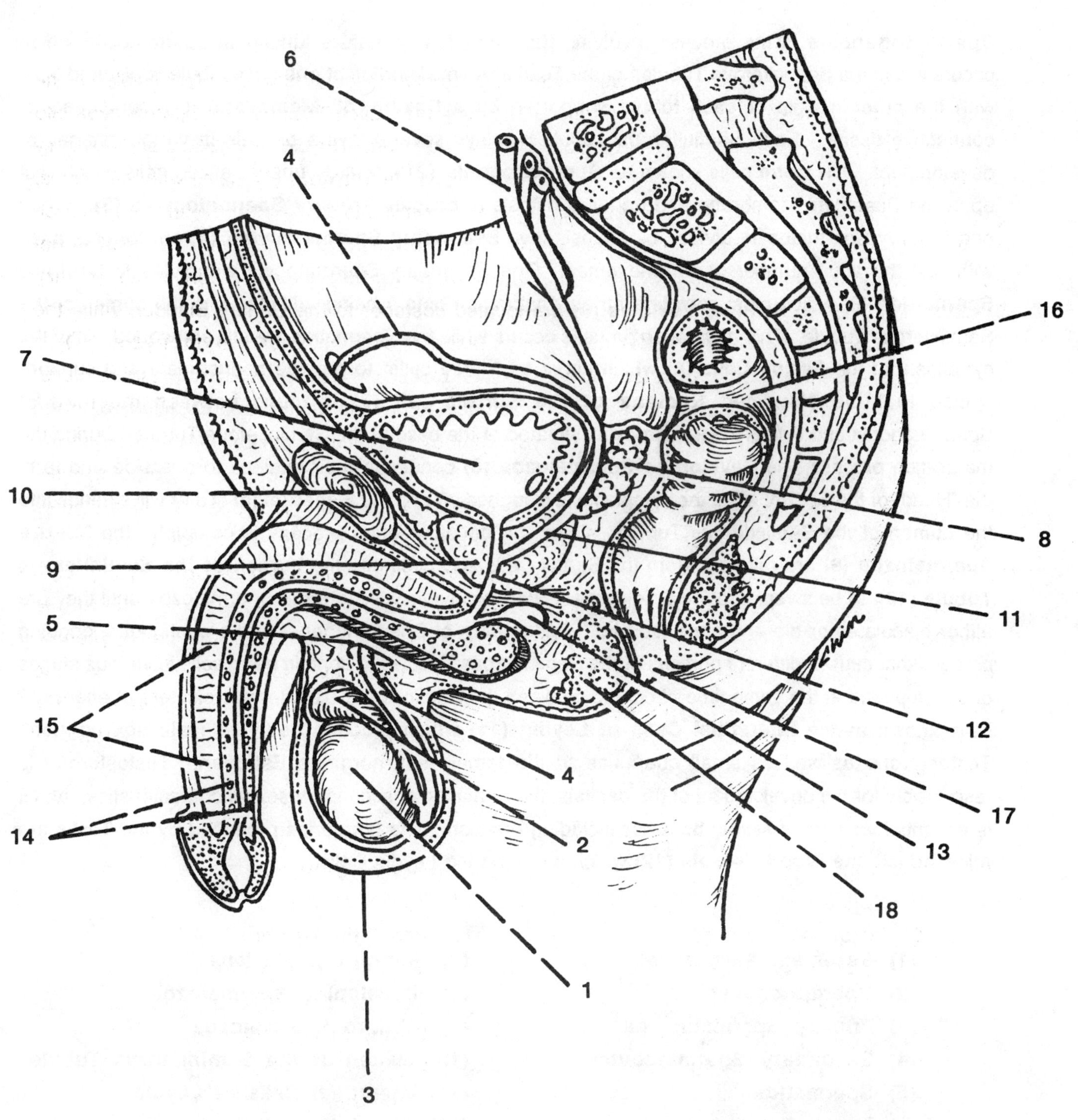

RP-1

Histology of the Testis

Spermatogenesis is the process involving the formation of mature functional spermatozoa which occurs within the Seminiferous Tubules of the Testis. A small portion of one such tubule is pictured here with the outer margin of each tubule supported by a **Basement Membrane (1)** composed of connective tissue. Each Seminiferous Tubule displays several layers of cells in various stages of development. The germ cells known as **Spermatogonia (2)** are large unspecialized cells which rest upon the Basement Membrane and undergo mitosis to become **Primary Spermatocytes (3)**. When one Primary Spermatocyte divides by Meiosis, two **Secondary Spermatocytes (4)** are formed, each with half the adult number of chromosomes. One Secondary Spermatocyte will divide to form two **Spermatids (5)** which must complete further maturation before being released into the Lumen of the Seminiferous Tubule. This maturation process occurs while the Spermatids are still embedded within the cytoplasm of the **Sertoli Cells (6)** which serve as "nurse cells" to provide nourishment and support. Notice that the elongated **Nucleus of the Sertoli Cell (7)** is found between the rows of Spermatogonia and Primary Spermatocytes situated at the base of the Seminiferous Tubule. During the maturation process, the **Developing Spermatozoa (8)** condense their nucleus to one side and form the "Head" of the Sperm, while the cytoplasm is sloughed off and the Flagellum (or "Tail") will extends into the Lumen of the Seminiferous Tubule. Once the developmental process is complete, the **Mature Spermatozoa (9)** are released from the Sertoli Cells and pass into the **Lumen of the Seminiferous Tubule (10)**, to be swept into the Epididymis. The Epididymis will store the Spermatozoa until they are either ejaculated or die and are absorbed. This pattern of Spermatogenesis is a continuous, ongoing process that causes different portions of the Seminiferous Tubules to contain germ cells in various stages of development at the same time. Located between the Seminiferous Tubules are clusters of endocrine cells known as the **Interstitial Cells of Leydig (11)** which produce both the male sex hormone Testosterone as well as small quantities of the female sex hormone Estrogen. Testosterone is responsible for the development of the genitals, the formation of secondary sexual characteristics, and it is essential for normal sexual behavior including erection. The hormones produced by the Testis are released into the **Blood Vessels (12)** which surround the Leydig Cells.

(1) Basement Membrane
(2) Spermatogonia
(3) Primary Spermatocytes
(4) Secondary Spermatocytes
(5) Spermatids
(6) Sertoli Cell Cytoplasm
(7) Sertoli Cell Nucleus
(8) Developing Spermatozoa
(9) Mature Spermatozoa
(10) Lumen of the Seminiferous Tubule
(11) Interstitial Cells of Leydig
(12) Blood Vessels

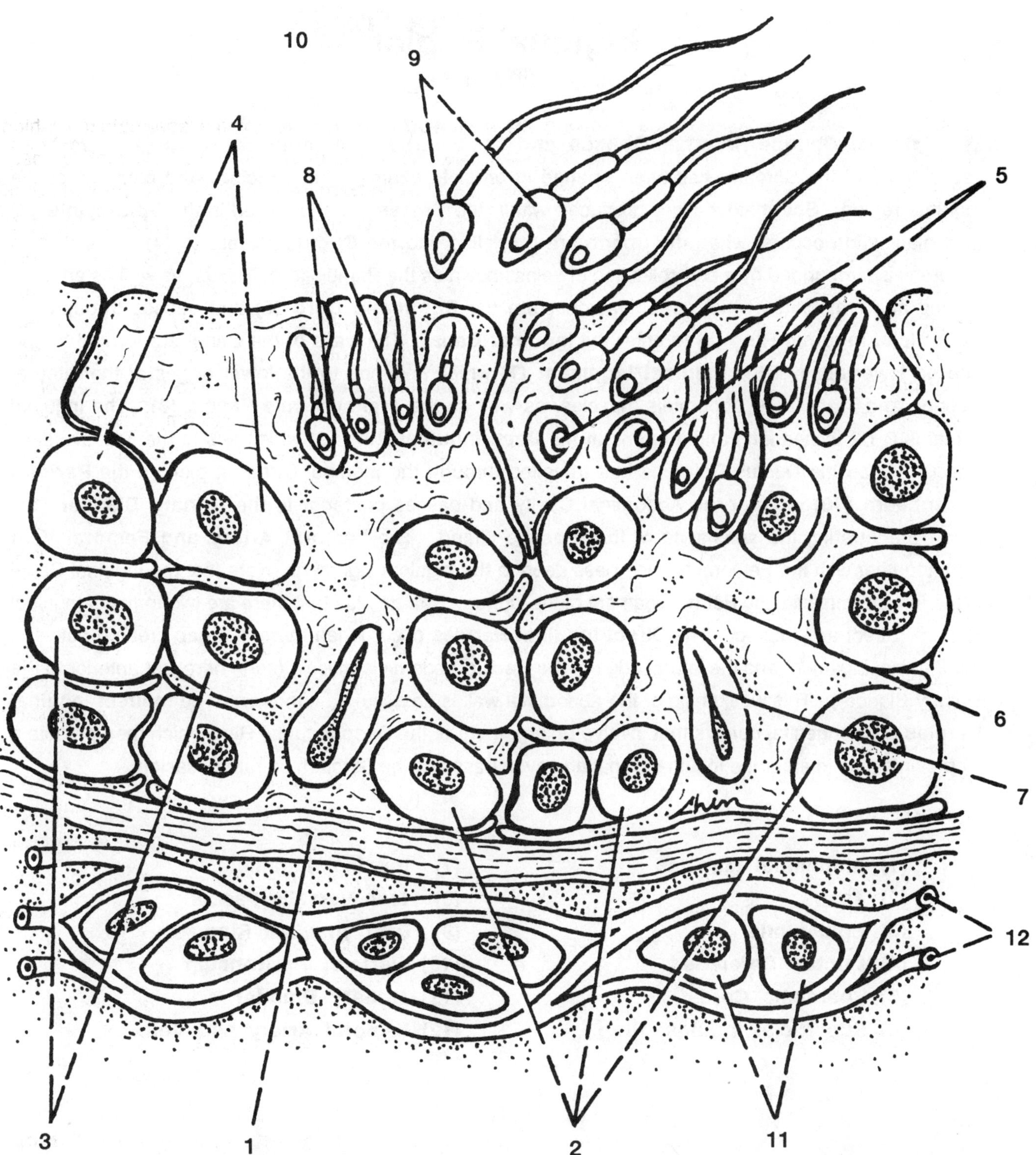
10
9
4
8
5
6
7
12
3
1
2
11

Inguinal Region
(Male)

The **External Oblique, Internal Oblique and Transverse Abdominis Muscles (1)** have been separated and the Scrotum has been opened in order to examine the structure and contents of the Inguinal region. Spermatozoa are produced within the **Testes (2)** and stored in the **Epididymis (3)** until ejaculation occurs, when the Sperm are propelled into the **Ductus Deferens (4)**. The Ductus Deferens is surrounded by a rich collection of veins known as the Pampiniform Plexus, as well as arteries, nerves and lymphatics which together constitute the **Spermatic Cord (5)** that passes through the **Superficial Inguinal Ring (6)** to enter the **Inguinal Canal**. The walls of the Canal are formed by the inrolled lower edge of the **Inguinal Ligament (7)** which is joined to the lower edges of the Internal Oblique and Transverse Abdominis muscles, as well as the **Transversalis Fascia (8)**. The Inguinal Canal is 2 1/2 inches long and it opens at the **Deep Inguinal Ring (9)** which lies in the Transversalis Fascia. Once the Ductus Deferens has travelled through the Inguinal Canal, it pierces the **Parietal Peritoneum (10)** to enter the Abdominal Cavity and passes posterior to the **Urinary Bladder (11)** before penetrating the substance of the Prostate Gland. The **Femoral Artery and Femoral Vein (12)**, together with the Femoral Nerve, pass deep to the Inguinal Ligament. Note that the Inguinal Canal in the female transmits the Round Ligament of the Uterus. Realize also that there are two forms of Inguinal Hernia: Direct and Indirect. The **Direct Inguinal Hernias** occur when there is a gap created between the Inguinal Ligament and the lateral edge of the Rectus Abdominis muscle (seen here just anterior to the Urinary Bladder). This weakening of the abdominal wall is an acquired condition. The **Indirect Inguinal Hernias** are congenital defects that involve enlargement of the Deep Inguinal Ring which then permits a loop of intestine to enter the Inguinal Canal and even pass into the Scrotum on rare occasion.

(1) Abdominal Muscles
(2) Testis
(3) Epididymis
(4) Ductus Deferens
(5) Spermatic Cord
(6) Superficial Inguinal Ring
(7) Inguinal Ligament
(8) Transversalis Fascia
(9) Deep Inguinal Ring
(10) Parietal Peritoneum
(11) Urinary Bladder
(12) Femoral Artery and Vein

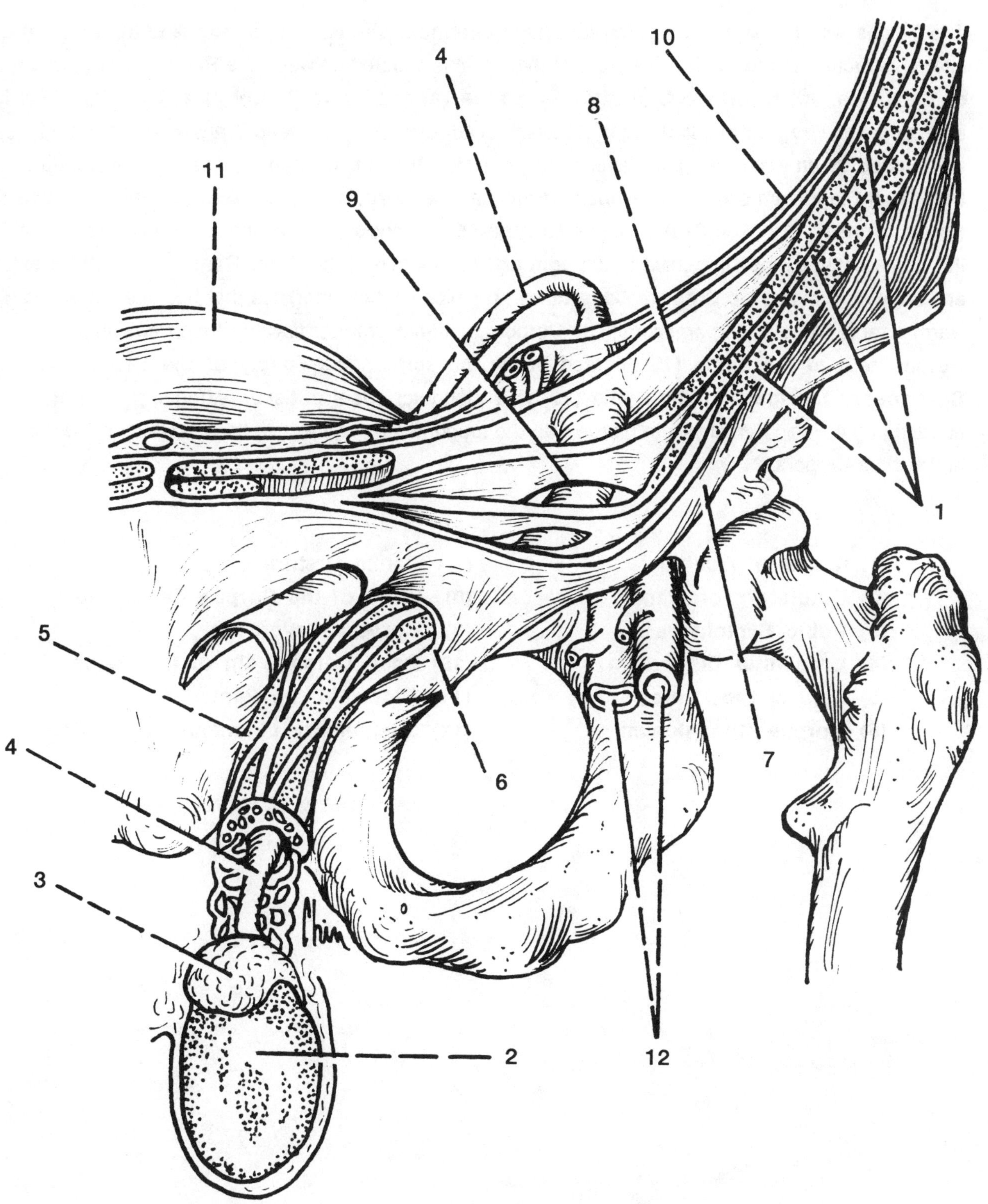

RP-3

External Male Genitalia

The Testes with the surrounding Scrotum have been removed in order to examine the external and internal structure of the Penis. The Root of the Penis is situated between the **Rami of the Ischium (1)** on each side, and below the **Obturator Foramina (2)** and **Pubic Symphysis (3)**. The Penis lies anterior to the **Urogenital Diaphragm (4)** which is composed of the Deep Transverse Perineal Muscle surrounded by an inner and outer layer of fascia. The **Bulb of the Penis (5)** is firmly attached to the Urogenital Diaphragm and it is continuous distally as the **Corpus Spongiosum (6)** which contains the **Penile Urethra (7)**. The **Crura of the Corpora Cavernosa (8)** are conical structures located at the angle between the Urogenital Diaphragm and the margin of the Pubic Ramus. The Crura extend anteriorly to form the two **Corpora Cavernosa (9)** which contain arterioles that become distended with warm arterial blood during erection. The Corpora, whether erect or flaccid, drain freely into the **Deep Dorsal Vein of the Penis (10)** located below the surrounding **Fascia of the Penis (11)**. The **Suspensory Ligament of the Penis (12)** is a triangular sheet of tissue extending from the under surface of the Pubic Symphysis to merge with the tough fibrous connective tissues which surround each of the three Corpora.

(1) Ramus of Ischium
(2) Obturator Foramen
(3) Pubic Symphysis
(4) Urogenital Diaphragm
(5) Bulb of the Penis
(6) Corpus Spongiosum
(7) Penile Urethra
(8) Crura of the Corpus Cavernosa
(9) Corpora Cavernosa
(10) Deep Dorsal Vein of the Penis
(11) Fascia of the Penis
(12) Suspensory Ligament of the Penis

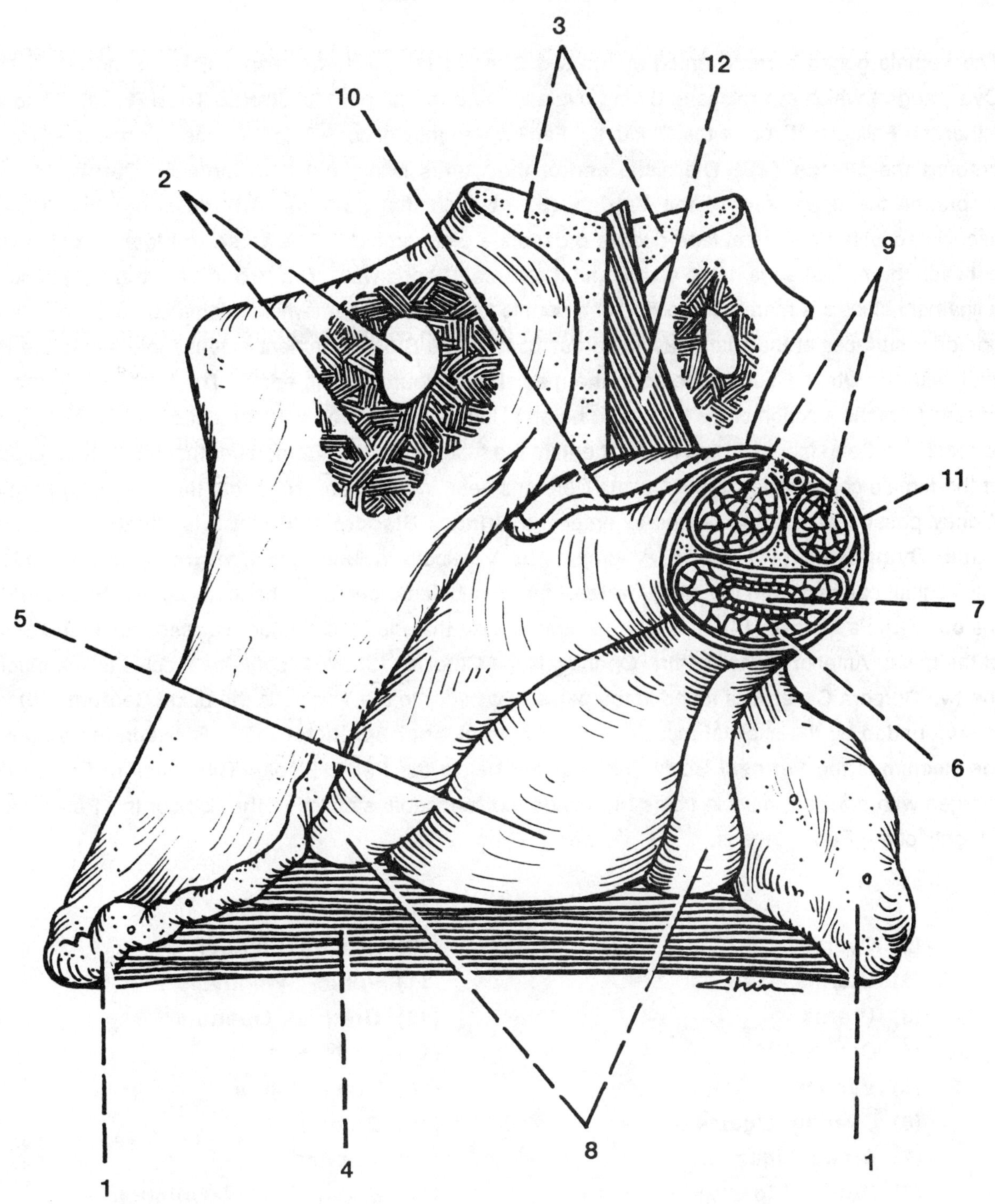
3
12
10
2
9
11
7
5
6
1
4
8
1
Chin

Female Pelvis
(Midsagittal View)

The Female gonad is represented by the two **Ovaries (1)** which are responsible for the production of Ova ("eggs") which are released during ovulation and swept into the **Uterine Tube (2)** (also known as either the Fallopian Tube or the Oviduct). Fertilization may occur within the tube or once the Ova have entered the **Uterus (3)**. The distal end of the Uterus is expanded to form the **Cervix (4)** which embraces the upper end of the **Vagina (5)**. Notice that portions of three supportive ligaments associated with the internal reproductive organs are pictured here. The **Ovarian Ligament (6)** extends between the medial edge of the Ovary and the lateral Uterine wall. The **Broad Ligament (7)** is not truly a ligament since it is merely composed of a double fold of peritoneum which originates on the anterior and posterior surfaces of the Uterus as its serous coats. The Broad Ligament extends laterally to the Pelvic wall, with the Uterine Tube situated in the medial three-fourths of its edge. The **Round Ligament (8)** passes from the junction of the Uterine Tube and Uterus, through the anterior leaf of the Broad Ligament, to reach the Deep Inguinal Ring where it enters the Inguinal Canal. Remember that this Round Ligament of the female corresponds to the Ductus Deferens seen in the male. Note that the **Ureter (9)** from each Kidney passes behind the Uterus to enter the **Urinary Bladder (10)** which is situated between the **Pubic Symphysis (11)** and the Vagina. The Vestibule includes the **Urethral Opening (12)** and the Vaginal opening, with the lateral walls of the Vestibule bordered by the inner **Labia Minora (13)** and the outer **Labia Majora (14)**. It is interesting to realize that the Labia Majora corresponds to the Scrotum of the male. Anterior to the Urethral Opening is the **Clitoris (15)** which contains erectile tissue much like the two Corpora Cavernosa found in the penis. Posterior to the Vagina is the distal **Rectum (16)** which is surrounded by the skeletal muscle of the **External Anal Sphincter (17)**. Between the Vagina and the Rectum is the **Perineal Body (18)** situated below the Pelvic Floor. This mass of fibrous tissue merges with adjacent muscle fibers to provide indispensable stability to the floor of the Pelvis and the integrity of the Pelvic Organs.

(1) Ovary
(2) Uterine Tube
(3) Uterus
(4) Cervix
(5) Vagina
(6) Ovarian Ligament
(7) Broad Ligament
(8) Round Ligament
(9) Ureter
(10) Urinary Bladder
(11) Pubic Symphysis
(12) Urethral Opening
(13) Labia Minora
(14) Labia Majora
(15) Clitoris
(16) Rectum
(17) External Anal Sphincter
(18) Perineal Body

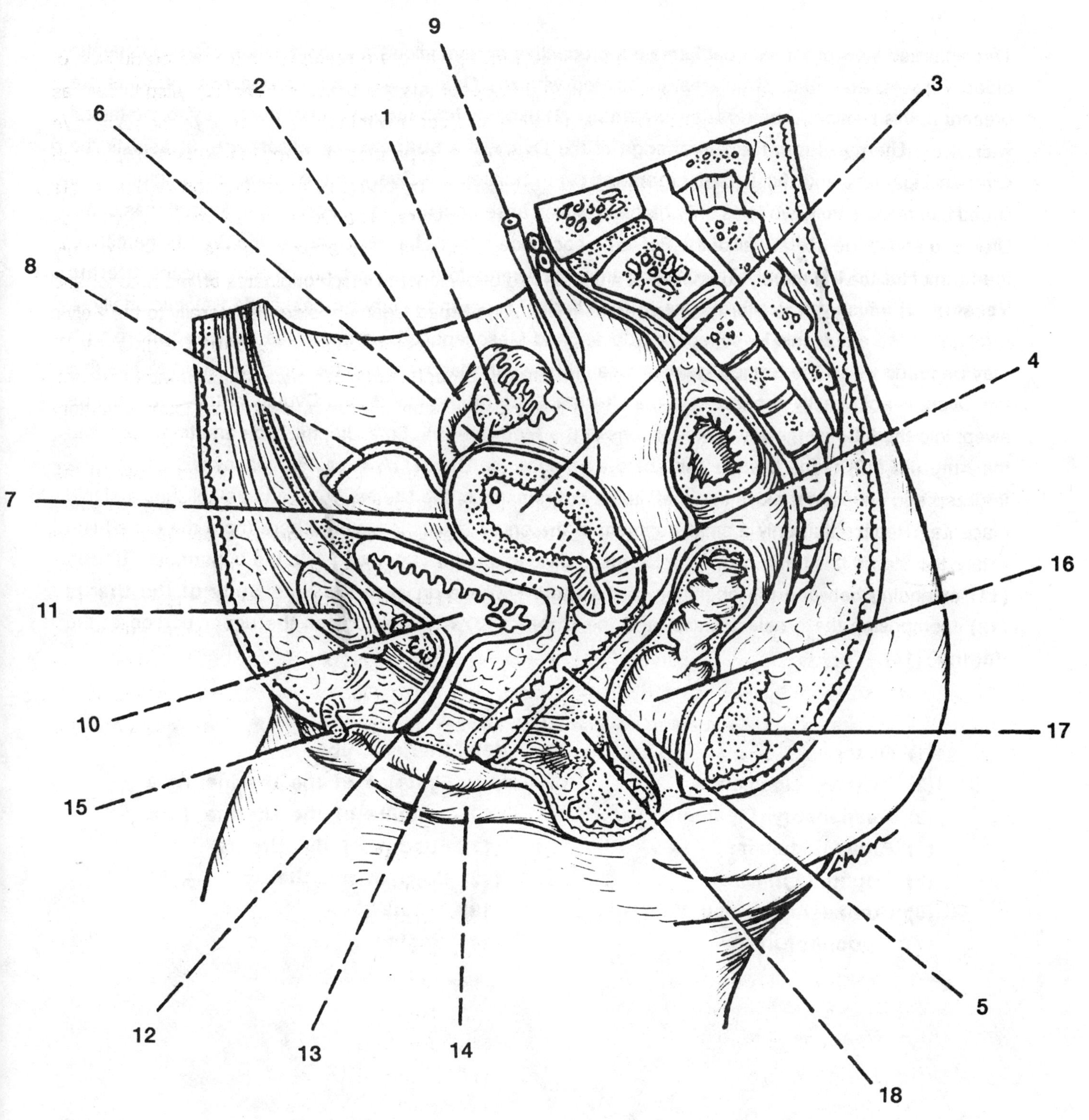
9
2
1
3
6
8
4
7
16
11
17
10
15
12
13
14
5
18

Structures Related to the Ovary

This enlarged view of the internal Female reproductive system allows an appreciation of the ligaments, blood vessels, and other structures associated with the **Ovary (1)**. Four supportive Ligaments are present in this section. The **Ovarian Ligament (2)** extends from medial wall of the Ovary to the lateral wall of the Uterus. From the lateral edge of the Ovary, the **Suspensory Ligament (3)** assists the Ovarian Ligament and the **Broad Ligament (4)** in holding the Ovary in the proper orientation. The Broad Ligament is truly two folds of Peritoneum which arise from the anterior and posterior surfaces of the Uterus to end at the lateral wall of the Pelvis on each side. Within the substance of the Broad Ligament is found the **Round Ligament (5)** extending from the Uterus to the Deep Inguinal Ring and the **Uterine Vessels (6)** which produce branches to the Uterine Tube and Ovary on each side. Occasionally, a remnant of the embryonic Mesonephric Tubules and Mesonephric Duct known as the **Epoophoron (7)** may be found within the Broad Ligament located between the Ovary and the Uterine Tube. Notice that the Ovary is not directly connected to the **Uterine Tube (8)**, so that during Ovulation, the Ova must be swept into the tube by the wave-like motions of the **Fimbriae (9)** of the Uterine Tube and then pass into the **Ampulla (10)** of the Uterine Tube before entering the Uterus. A Tubal Pregnancy will occur if the fertilized egg should not move into the Uterus but remains in the Uterine Tube and establishes a viable placenta. Realize that only a small segment of the entire Uterus is pictured here (note the dotted line where the uterus has been cut). The Uterus is divided into three portions with the uppermost **Fundus (11)** extending between the openings of the Uterine Tubes on each side, the **Body of the Uterus (12)** comprising the greatest mass, and the **Cervix (13)** which embraces the upper portion of the **Vagina (14)**.

(1) Ovary
(2) Ovarian Ligament
(3) Suspensory Ligament
(4) Broad Ligament
(5) Round Ligament
(6) Uterine Artery and Vein
(7) Epoophoron
(8) Uterine Tube
(9) Fimbriae of the Uterine Tube
(10) Ampulla of the Uterine Tube
(11) Fundus of the Uterus
(12) Body of the Uterus
(13) Cervix
(14) Vagina

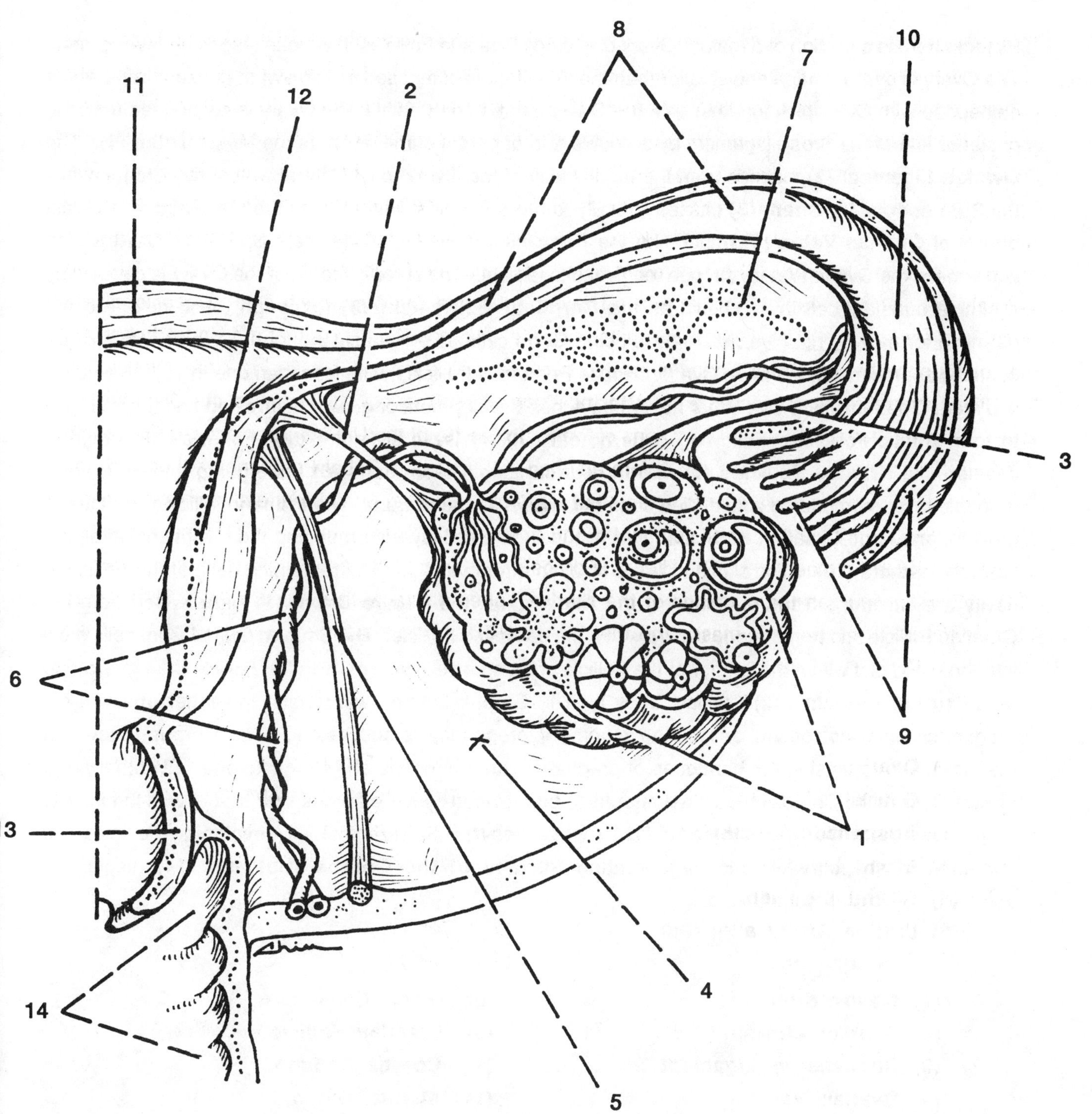
8
10
7
11
12
2
3
6
9
13
1
4
14
5

Histology of the Ovary

Pictured here is a section of a mature Ovary displaying Ova and Follicles in various stages of development. The Ovary is oval in shape and smaller than the Testis. It is composed of somewhat dense fibrous tissue embedded with Ova which together give the Ovary a firm consistency. The Ovary is suspended from the posterior leaf of the Broad Ligament by a double fold of peritoneum known as the **Mesovarium (1)**. The **Ovarian Ligament (2)** extends from the medial wall of the Ovary to the lateral wall of the Uterus, while the **Suspensory Ligament (3)** passes laterally to the abdominal wall. Within the Mesovarium is a vast plexus of **Ovarian Veins (4)** (much like the Pampiniform Plexus of the male Spermatic Cord) which accompany the Ovarian Artery through the Broad Ligament. The fibrous stroma of the Ovary is covered by a simple cuboidal cell layer known as the **Germinal Epithelium (5)** (although these cells are not germinal for they do not give rise to developing Germ cells as previously thought). At birth, each ovary contains approximately 3 million Ova (known as Primordial Follicles) with less than one in 15,000 actually ovulated. The remaining Ova either never begin their development or growth is arrested and they atrophy to form Atretic Follicles. Following puberty, each month several **Primordial Follicles (6)** begin to develop into **Primary Follicles (7)** with the surrounding **Thecal cells (8)** undergoing mitosis to form a double layer surrounding a fluid-filled chamber known as the Antrum. The Primary Follicles continue to grow to form large **Graafian Follicles (9)** which include a cell layer surrounding the Ovum known as the **Corona Radiata (10)** which clings to the Ova even after ovulation. The remaining cells of the Follicular cavity are called the Stratum Granulosum. At ovulation, the **Mature Ovum (11)** is released from the Graafian Follicle and hemorrhage occurs within the collapsed Follicle. The Stratum Granulosum cells grow into the blood clot within the interior of the Follicle creating a yellow color, with the Follicle then referred to as a **Corpus Luteum (12)** ("yellow body"). The Corpus Luteum will persist for either one week if pregnancy does not occur, or for nine months if pregnancy is successful. The Corpus Luteum is responsible for releasing the hormones of pregnancy, the Estrogens and Progesterone. Eventually, the Follicle Cells of the Corpus Luteum will atrophy to be replaced by white fibrous scar tissue, with the Follicle then known as a **Corpus Albicans (13)** ("white body"). Realize that all developing Follicles are spherical in shape, while the degenerating **Surface of the Corpus Albicans (14)** displays a complex, convoluted appearance.

(1) Mesovarium
(2) Ovarian Ligament
(3) Suspensory Ligament
(4) Ovarian Veins
(5) Germinal Epithelium
(6) Primordial Follicle
(7) Primary Follicles
(8) Thecal Cells
(9) Graafian Follicle
(10) Corona Radiata
(11) Mature Ovum
(12) Corpus Luteum
(13) Corpus Albicans
(14) Surface of Corpus Albicans

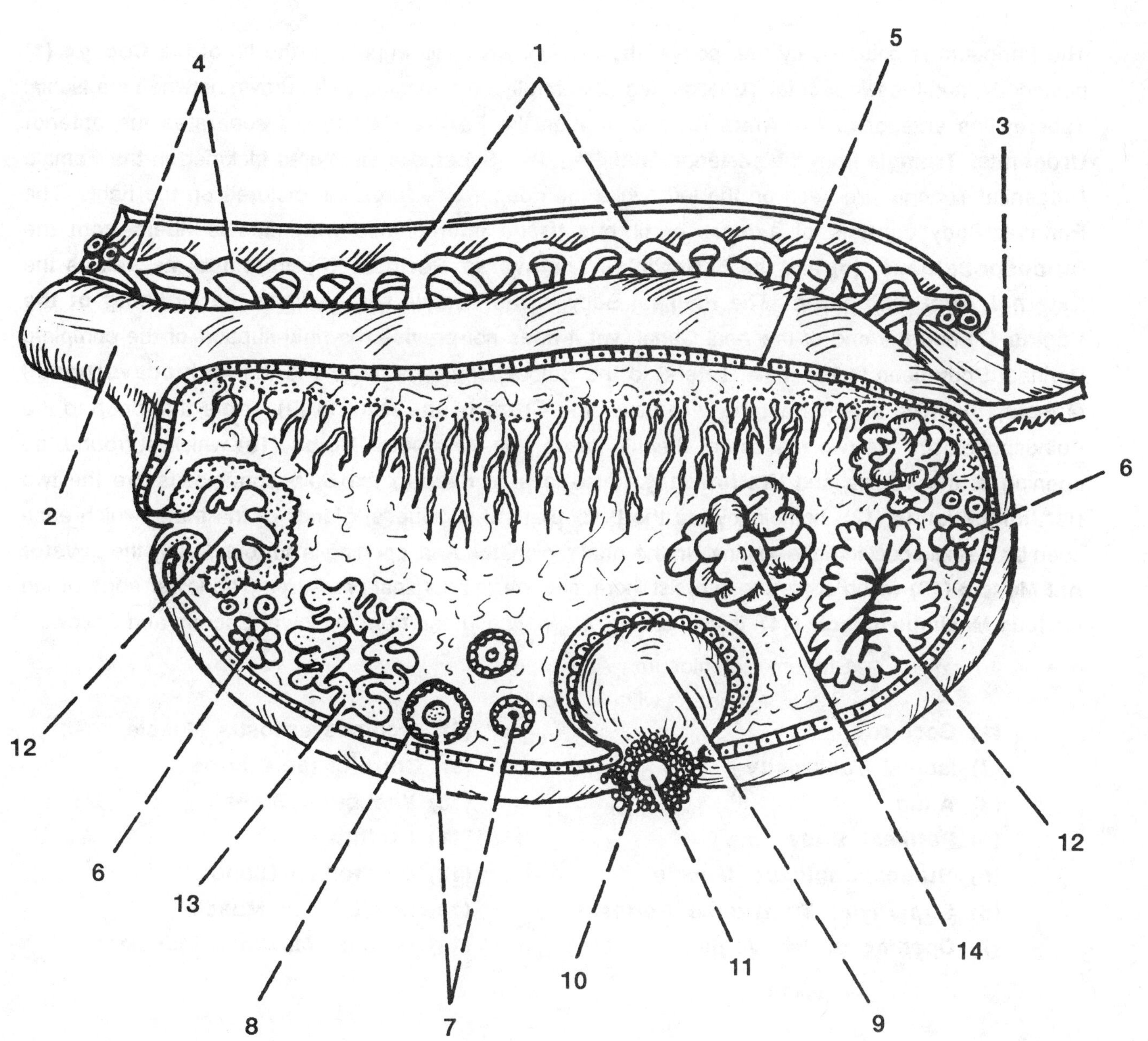
1
2
3
4
5
6
6
7
8
9
10
11
12
12
13
14
Chin

Female Perineum

The Perineum is bounded by four points: the Pubic Symphysis anteriorly, the tip of the **Coccyx (1)** posteriorly, and the two **Ischial Tuberosities (2)** laterally. An imaginary line drawn between the Ischial Tuberosities anterior to the **Anus (3)** and through the **Perineal Body (4)** separates the anterior **Urogenital Triangle** from the posterior **Anal Region**. Superficial structures included in the Female Urogenital Triangle are seen on the left, while the deeper structures are pictured on the right. The Perineal Body consists of a mass of fibrous tissue intermingled with muscle fibers from the **Bulbospongiosus (5)** and the **Superficial Transverse Perineal (6) muscles**, as well as the **External Anal Sphincter**. The Perineal Body in the Female lies between the **Opening of the Vagina (7)** and the end of the Anal Canal, yet it does not provide the rigid support of the complete Perineal Body seen in the male. Lateral to the Bulbospongiosus muscle is the **Ischiocavernosus Muscle (8)** which extends anteriorly to cover the **Crura of the Clitoris (9)**. Note that deep to the Bulbospongiosus are two masses of erectile tissue, the **Vestibular Bulbs (10)** which surround the opening of the Vagina and **Urethra (11)**. Also located deep to the Bulbospongiosus are the two **Bartholin Glands (12)**, homologous to the Bulbourethral (Cowper's) glands of the male, which each open by a single duct into the Vagina. In the Anal region, the Anal opening is surrounded by the **Levator Ani Muscle (13)** which contracts to resist increased intra-abdominal pressure. The lower edge of the **Gluteus Maximus muscle (14)** is also shown, originating in part from the lower Sacrum and Coccyx.

(1) Coccyx
(2) Ischial Tuberosity
(3) Anus
(4) Perineal Body
(5) Bulbospongiosus Muscle
(6) Superficial Transverse Perineal
(7) Opening of the Vagina
(8) Ischiocavernosus Muscle
(9) Crura of the Clitoris
(10) Vestibular Bulbs
(11) Urethra
(12) Bartholin's Gland
(13) Levator Ani Muscle
(14) Gluteus Maximus Muscle

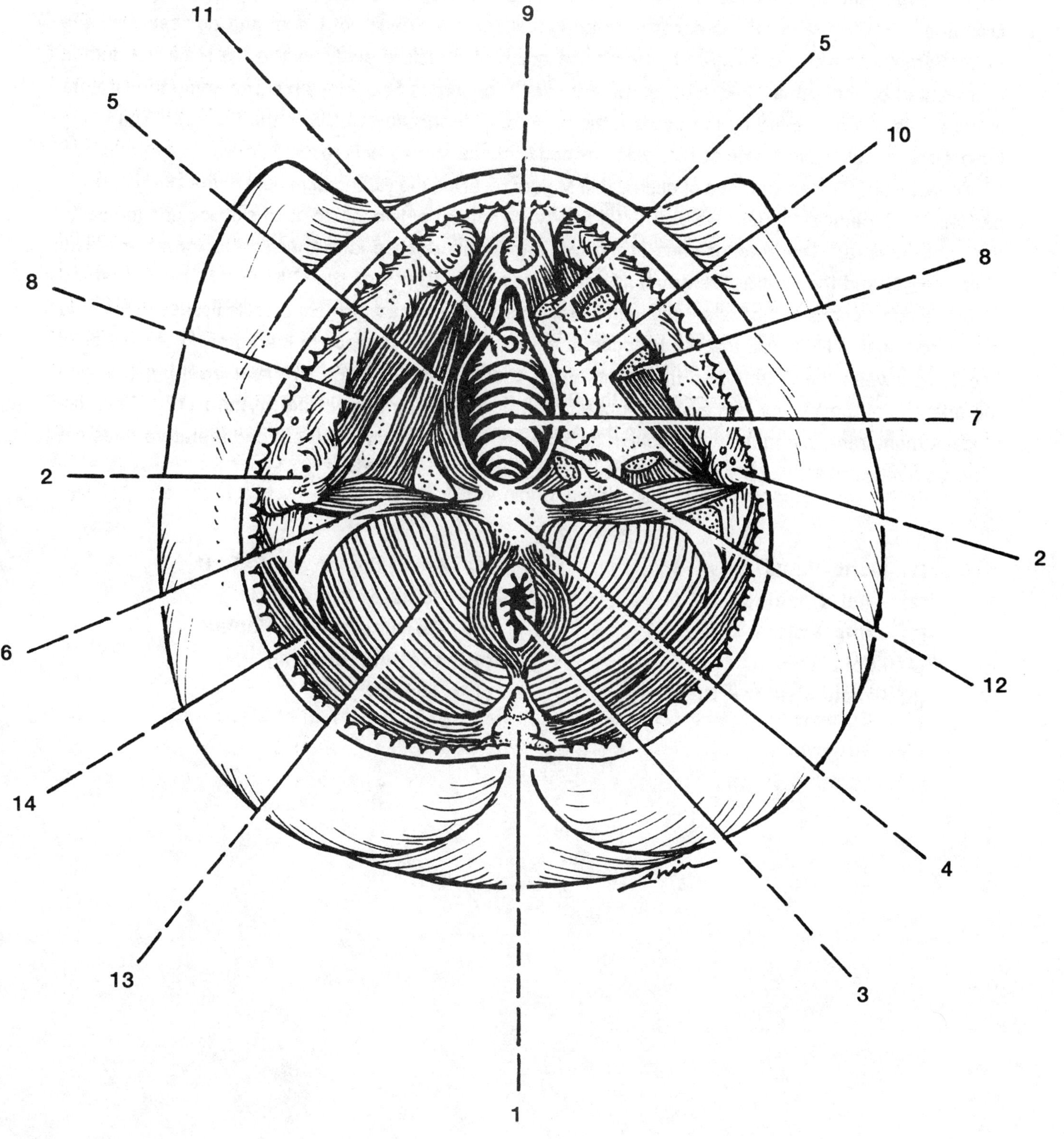

RP-8

External Female Genitalia

The External Female Genitalia is located posterior to the **Mons Pubis (1)** and anterior to the **Anal Opening (2)**. The Mons Pubis consists of an elevation or eminence of fascia and adipose over the Pubic Symphysis which is covered by hair in the adult. A structure equivalent to the Male Scrotum is represented by the **Labia Majora (3)** of the Female. The Corpus Spongiosum surrounding the Urethra in the Penis is represented by the **Labia Minora (4)** and **Vestibular Bulbs** of the Female. The Labia Majora are two elongated folds of skin which extend from the Mons Pubis to the Perineal Body (indicated by a doted line). The anterior end of each Labia Majora receives the Round Ligament of the Uterus once it has left the Superficial Inguinal Ring. The Labia Minora consist of two thin folds of skin located medial to the Labia Majora. The anterior ends of the two Labia Minora fuse in the midline to form the dorsal **Prepuce (5)** and the ventral **Frenulum (6)** surrounding the **Glans of the Clitoris (7)**. Note that the Clitoris is homologous to the Male Penis with two Corpora Cavernosa which are erectile tissues in both the male and female. Note also that the Prepuce of the Clitoris is homologous to the Foreskin of the Penis. Located between the Clitoris and the **Opening of the Vagina (8)** is the **Urethral opening (9)**, with the opening of the Vagina partially surrounded by the delicate remains of the **Hymen (10)**. This thin mucous membrane is frequently variable in its form and extent, easily ruptured, and therefore does not serve as a reliable indicator of virginity, as previously believed.

(1) Mons Pubis
(2) Anal Opening
(3) Labia Majora
(4) Labia Minora
(5) Prepuce of the Clitoris
(6) Frenulum of the Clitoris
(7) Glans of the Clitoris
(8) Opening of the Vagina
(9) Urethral opening
(10) Hymen

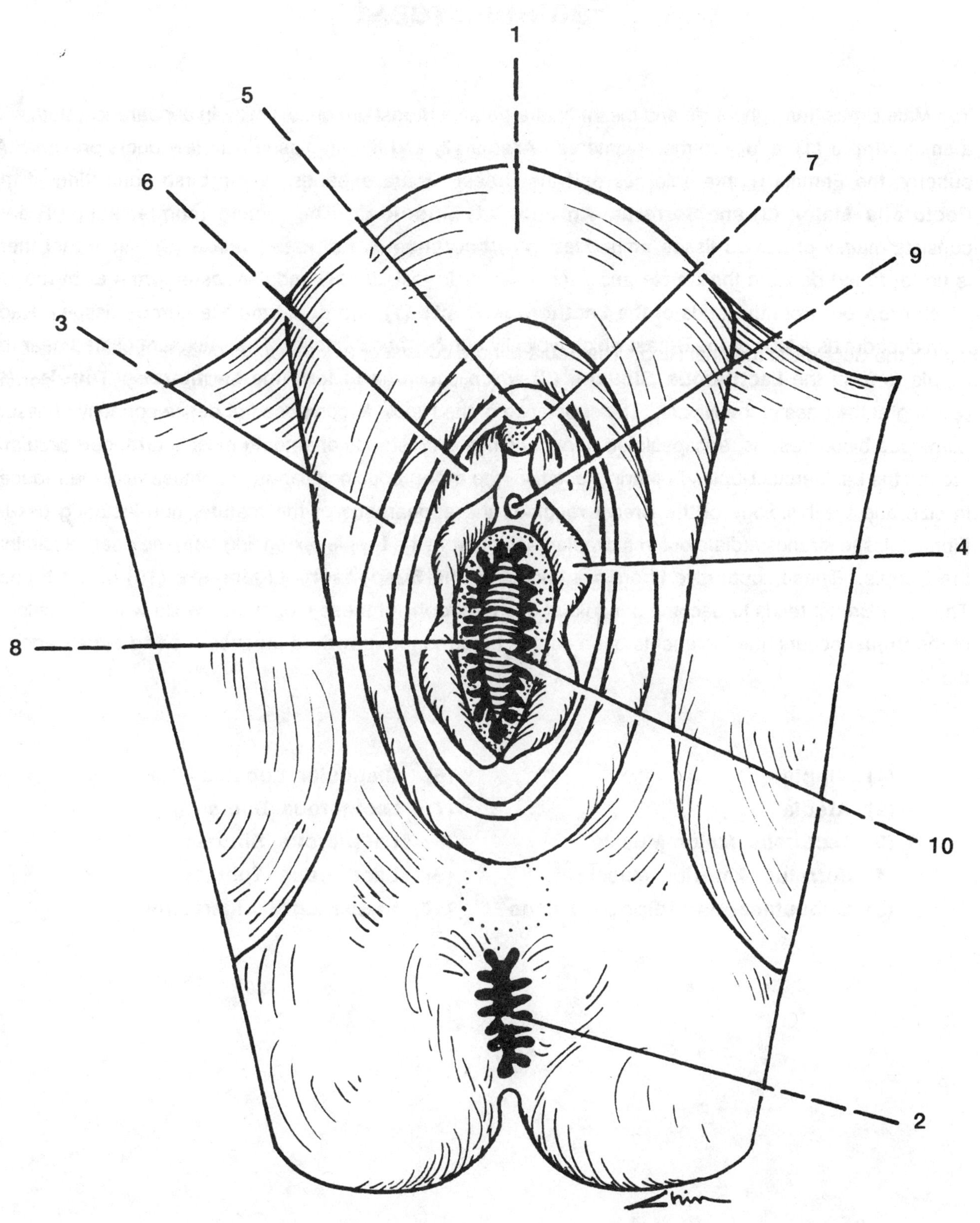
1
5
7
6
9
3
4
8
10
2

Female Breast

The Male Breast throughout life and the immature Female Breast are quite similar in appearance, both with a small **Nipple (1)**, a fully formed pigmented **Areola (2)** and fibrous tissue with few ducts present. At puberty, the Female Nipple enlarges and the Breast Tissue expands over a base consisting of the **Pectoralis Major (3)** and **Serratus Anterior (4) Muscles**. The resting (non-lactating) Breast consists mainly of fibrous tissue embedded in **Subcutaneous Adipose Tissue (5)**. Note that there is no fat found beneath the Nipple and Areola. Prior to lactation, Glandular tissue forms **Lobules (6)** which grow out from the ends of the **Lactiferous Ducts (7)** and penetrate the fibrous tissue. Each main duct drains a lobe of the Breast which typically number about fifteen. The Ducts are dilated near the Nipple to form the **Lactiferous Sinuses (8)** which converge to form the **Lactiferous Tubules (9)** opening at the base of the Nipple. The substance of the Nipple is composed of dense connective tissue, numerous blood vessels, encapsulated nerve endings, with stands of smooth muscle arranged circularly around the Lactiferous Ducts. Following lactation, the milk-producing Glandular Lobules become reduced in size and the histology of the Breast resumes the appearance of the mature, non-lactating tissue. Fibrous tissue strands radiate out in a circular fashion from the Nipple, extending from the deep fascia into the Dermis. These supportive fibers are known as the **Suspensory Ligaments (10)** of the breast. The older Breast tends to become pendulous due to atrophy of these Ligaments, while when Carcinoma of the Breast occurs, the Ligaments often contract around the growth to produce a pitted appearance in the skin.

(1) Nipple
(2) Areola
(3) Pectoralis Major Muscle
(4) Serratus Anterior Muscle
(5) Subcutaneous Adipose Tissue
(6) Glandular Lobules
(7) Lactiferous Ducts
(8) Lactiferous Sinuses
(9) Lactiferous Tubules
(10) Suspensory Ligaments

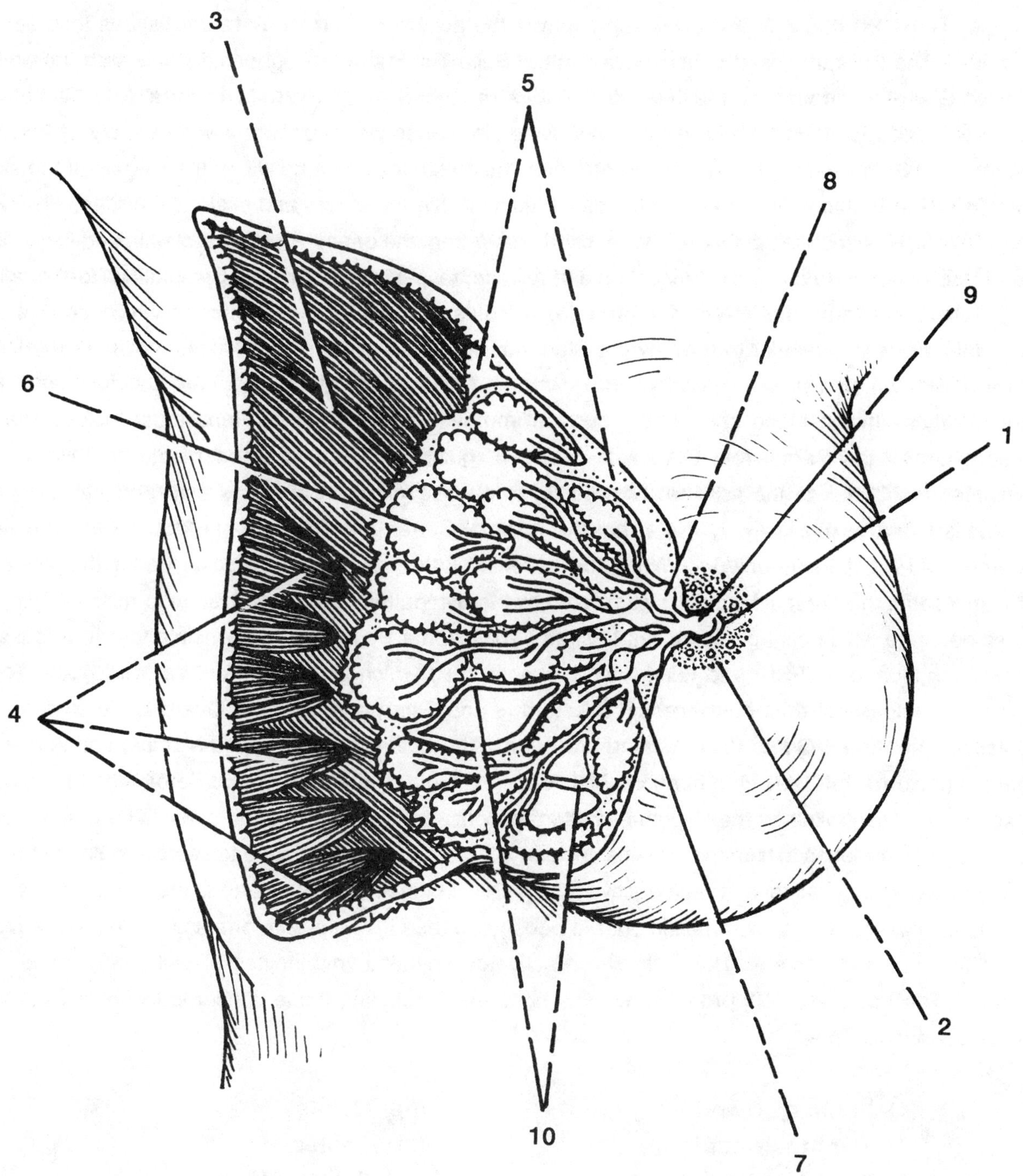

RP-10

Endocrine Gland Locations

The various glands within the body either secrete their products into ducts or deposit their secretions directly into the bloodstream. The former are known as Exocrine Glands which include the Tear Glands, Sweat Glands, as well as Digestive, Pancreatic and Liver Glands. The ductless Endocrine Glands produce Hormones that enter the general circulation to execute their effects upon target organs located elsewhere. The so-called "Master Gland" of the body is the **Pituitary Gland (1)** which controls the activity of numerous target tissues located throughout the body. The Pituitary Gland is housed within the Sella Turcica of the Sphenoid Bone, with the activity of this "Master Gland" controlled by the releasing factors produced in the **Hypothalamus (2)**. Stimulation of the Pituitary Gland results in the release of some Nine Hormones which control a wide variety of body functions. Melanocytes Stimulating Hormone (MSH) activates the melanocytes found within the Dermis of the skin, Growth Hormone (GH) promotes expansion of the tissue mass within the bones and skeletal muscles, while Antidiuretic Hormone (ADH) stimulates the Kidney to reabsorb water from the urine. Pituitary Prolactin and Oxytocin act upon the Breast Tissue promoting milk production and release, respectively. Thyroid Stimulating Hormone (TSH) from the Pituitary activates the **Thyroid Gland (3)** to release the Thyroid Hormones which control the rate of metabolic activity. Posterior to the Thyroid Gland are four distinct islands of tissue known as the **Parathyroid Glands (4)** which release Parathyroid Hormone (PTH) serving to regulate Calcium and Phosphate concentrations in the blood and bone. Surgical removal of the Thyroid Gland must include the deliberate preservation of the Parathyroid Glands or death will quickly follow. Adrenocorticotrophic Hormone (ACTH) stimulates the Cortex of the **Adrenal Glands (5)** to release various Corticoids that regulate electrolytes, food metabolism and sexual activity. Note that the Adrenal Cortex and the Adrenal Medulla are developmentally different and functionally unique. While the Adrenal Cortex releases Corticoids under the influence of the Pituitary Gland, the Adrenal Medulla is stimulated by the Sympathetic Nervous System to release Epinephrine and Norepinephrine which control responses to stress. Lastly, the Pituitary is responsible for the release of Follicle Stimulating Hormone (FSH) and Luteinizing Hormone (LH) which act upon the **Ovaries (6)** and **Testes (7)** to control the release of their hormones, Progesterone and Estrogens from the Ovary or Testosterone from the Testes. The **Pineal Gland (8)** is located posterior to the Thalamus and Hypothalamus, and is responsible for both the nocturnal release of the hormone Melatonin and the daytime secretion of Serotonin. This diurnal rhythm of secretion is controlled by the Hypothalamus in response to light entering the eyes. The precise role of these Pineal Gland hormones in humans remains unclear, although it has been found to secrete Melatonin and may play a regulatory role in sexual and reproductive functions. Located over the Heart is the **Thymus Gland (9)** which obtains its maximum size and function prior to puberty, to then involute and form scar tissue in the mature adult. The Thymus Gland secretes Thymosin which activates Thymic Lymphocytes (T-cells) originating in the Bone Marrow. The **Pancreas (10)** produces several hormones, including those concerned with the control of blood glucose concentrations.

(1) Pituitary Gland
(2) Hypothalamus
(3) Thyroid Gland
(4) Parathyroid Glands
(5) Adrenal Glands
(6) Ovaries
(7) Testes
(8) Pineal Gland
(9) Thymus Gland
(10) Pancreas

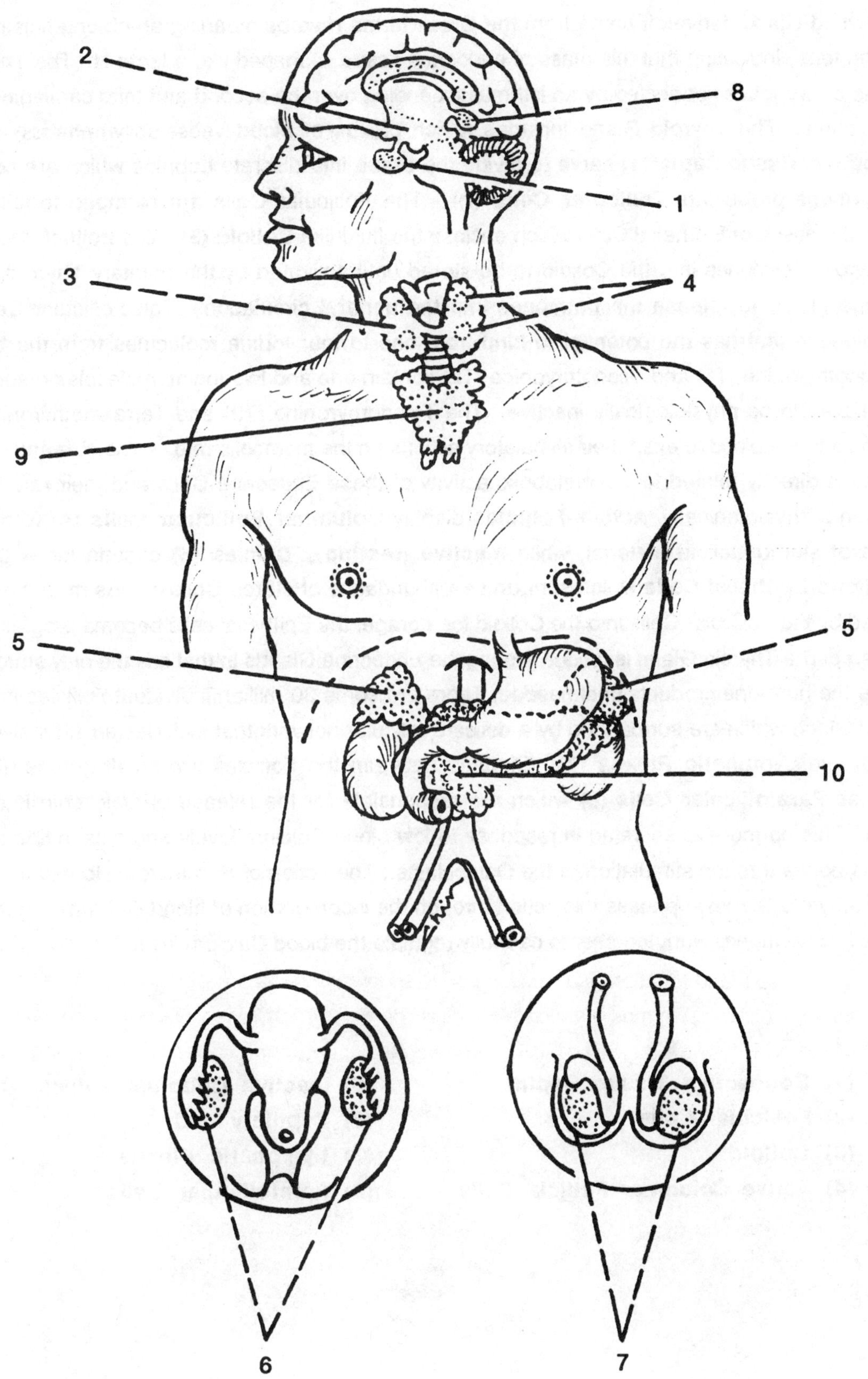

ES-1

Histology of the Thyroid Gland

The Thyroid Gland derives it name from the Greek terms *thyreos* meaning an oblong shield and *eidos* meaning form, indicating that this mass of endocrine tissue is shaped like a large H. The Thyroid Gland consists of two lobes connected by an isthmus extending over the second and third cartilaginous rings of the Trachea. The Thyroid Gland includes a rich plexus of blood vessels, lymphatics and nerves. **Connective Tissue Septa (1)** serve to divide the tissue into discrete Lobules which are composed of the hormone-producing **Follicular Cells (2)**. The Follicular Cells are clumped together to form spherical masses of Epithelial Cells which enclose the fluid-like **Colloid (3)**. The Epithelial Cells secrete the Thyroid Hormones into the Colloid to be stored until activated by the Pituitary Thyroid Stimulating Hormone (TSH) to release the hormones into the general circulation. The Follicular Cells secrete Thyroglobulin that has the potential to bind from one to four Iodine molecules from the blood. The Monoiodothyronine (T1) and Diiodothryonine (T2) contain one and two Iodine molecules respectively and are believed to be physiologically inactive. The Triiodothyronine (T3) and Tetraiodothyronine (T4) are secreted into the blood to exert their stimulatory actions on the metabolic rate. The size and shape of the Follicles is directly related to the metabolic activity of these Endocrine Cells and their rate of hormone secretion. Physiologically **Active Follicles** display **Columnar Follicular Cells (4)** with a reduced volume of stored Colloid material, while **Inactive (resting) Follicles (5)** contain more **Cuboidal** to **Squamous Epithelial Cells** which surround an abundance of stored Colloid. (As more hormones are secreted by the Follicular Cells into the Colloid for storage, the Epithelial cells become progressively more flattened.) The Thyroid Gland is unique among the Endocrine Glands in that it is the only structure known to store the hormone products until needed. There are some 30 million individual Follicles in the human Thyroid Gland which are surrounded by a delicate reticular network that includes an extensive **Capillary bed (6)** and **Lymphatic Plexus (7)**. Situated between the Follicles are small groups of pale cells known as **Parafollicular Cells (8)** which are responsible for the release of Calcitonin into the blood stream. This hormone is secreted in response to low blood Calcium levels and acts to liberate Calcium from the bones through stimulation of the Osteoclasts. The action of Parathyroid Hormone (secreted by the Parathyroid Glands) opposes this action through the incorporation of blood Calcium into the bone, so that the two hormones work together to carefully regulate the blood Calcium levels.

(1) Connective Tissue Septa
(2) Follicular Cells
(3) Colloid
(4) Active Columnar Follicle Cells
(5) Inactive Cuboidal Follicle Cells
(6) Capillary bed
(7) Lymphatic Plexus
(8) Parafollicular Cells

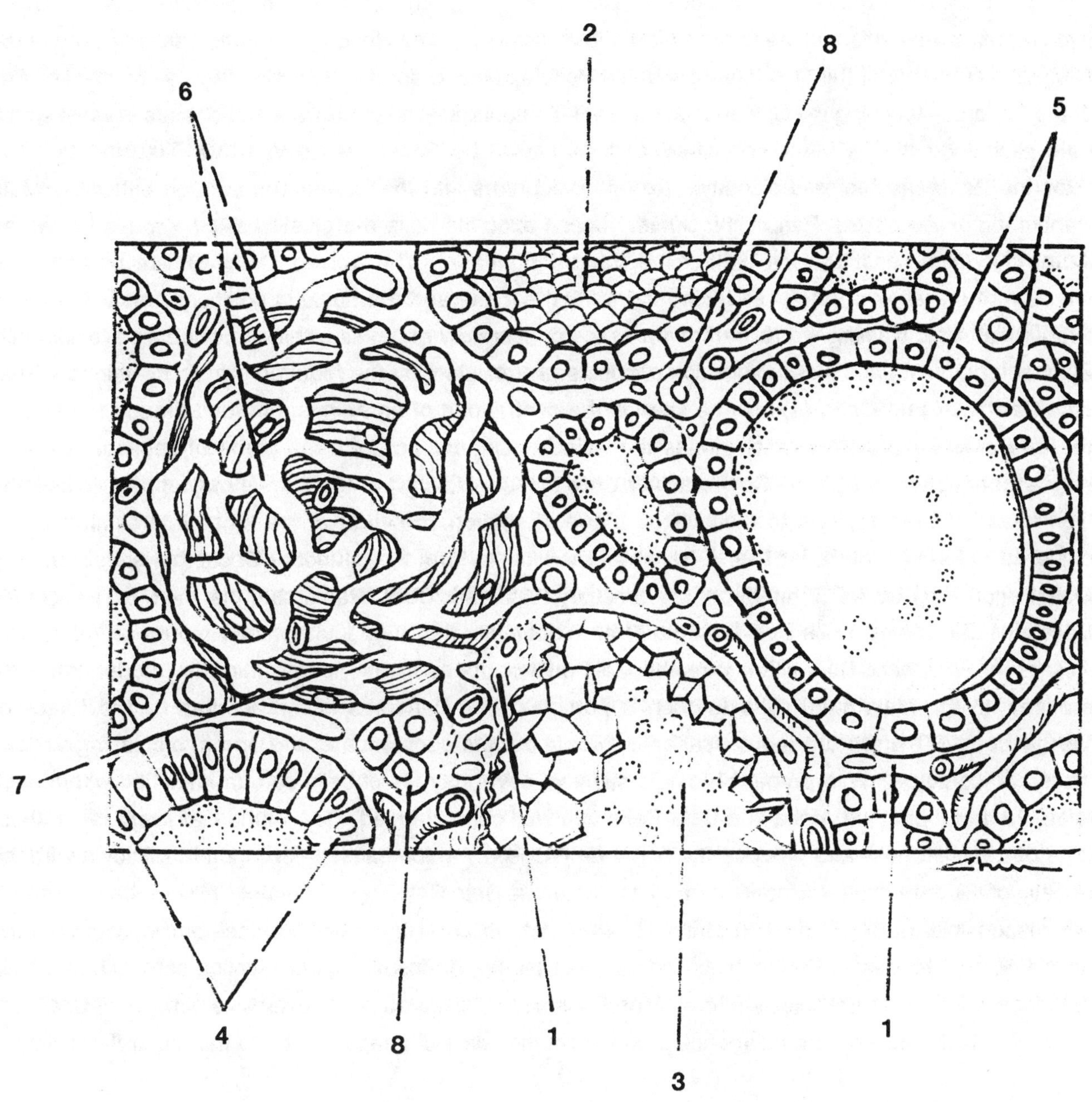

ES-2

Histology of the Pancreas

As noted earlier, the Pancreas has a dual function, being both an Exocrine and Endocrine Gland. These two purposes are evident in the examination of the Histology. The Head of the Pancreas is nestled in the C-shaped curvature of the Duodenum, with the Body extending across the abdominal cavity, and the Tail of the Pancreas touching the Spleen in the upper left quadrant. The fresh Pancreatic tissue is white with a pale pink tinge that is highly lobulated and separated by **Septa of Connective Tissue (1)**. The Exocrine Pancreas secretes digestive enzyme precursors into the Duodenum through either the Main Pancreatic or Accessory Pancreatic Ducts. These exocrine cells are clustered into groups known as **Acini (2)** which constitute the bulk of the Pancreatic mass. The **Acinar Cells (3)** are pyramidal in shape with the apex of each cell facing the central ductule and the nucleus positioned near the base. Proper histologic staining will reveal the presence of secretory granules within the cytoplasm (acidophilic Zymogen granules) and extensive rough endoplasmic reticulum for the required protein synthesis. Small pale **Centro-acinar Cells (4)** may be seen in the central part of an Acinus, grouped around the start of the **Intercalated Duct (5)** which drains the Acinus. Depending upon the plane of section, the Acini may appear spherical with the Centro-acinar Cells in the middle (4) or the Acinus may appear to lack the Centro-acinal Cells (5) due to the oblique plane of section. Several of the small Intercalated Ducts coalesce to form **Intralobular Ducts (6)** which then join to form the Interlobar Ducts that either form the Main Pancreatic Duct (of Wirsung) or the Accessory Pancreatic Duct. Remember that the Main Pancreatic Duct joins the Common Bile Duct from the Liver to open into the Duodenum at the Ampulla of Vater. The Accessory Pancreatic Duct either joins the Main duct or opens independently into the Duodenum, 2 cm proximal to the Main duct. The Endocrine portion of the Pancreas is represented by the **Islets of Langerhans (7)** which are found scattered throughout the substance of the organ, displaying various sizes and contours. With proper histologic staining, several types of cells can be identified within each Islet. Two of these, the Alpha and Beta Cells, are well characterized. The Alpha Cells represent about 20% of the Islet mass and produce the hormone Glucagon which causes Glycogen breakdown with the release of Glucose into the bloodstream. The numerous Beta Cells (approximately 75% of the Islet mass) are responsible for the production of Insulin which acts to lower the blood Glucose and encourages the formation of Glycogen within the Liver and Muscle tissues. An insufficient production of Insulin results in the disease known as Diabetes Mellitus. The Pancreas is infiltrated by an extensive network of **Capillary beds (8)** which transport the Hormones produced in the Islets of Langerhans into the general circulation.

(1) Connective Tissue Septa
(2) Acini
(3) Acinar Cells
(4) Centro-acinar Cells
(5) Intercalated Ducts
(6) Intralobular Ducts
(7) Islets of Langerhans
(8) Capillary beds

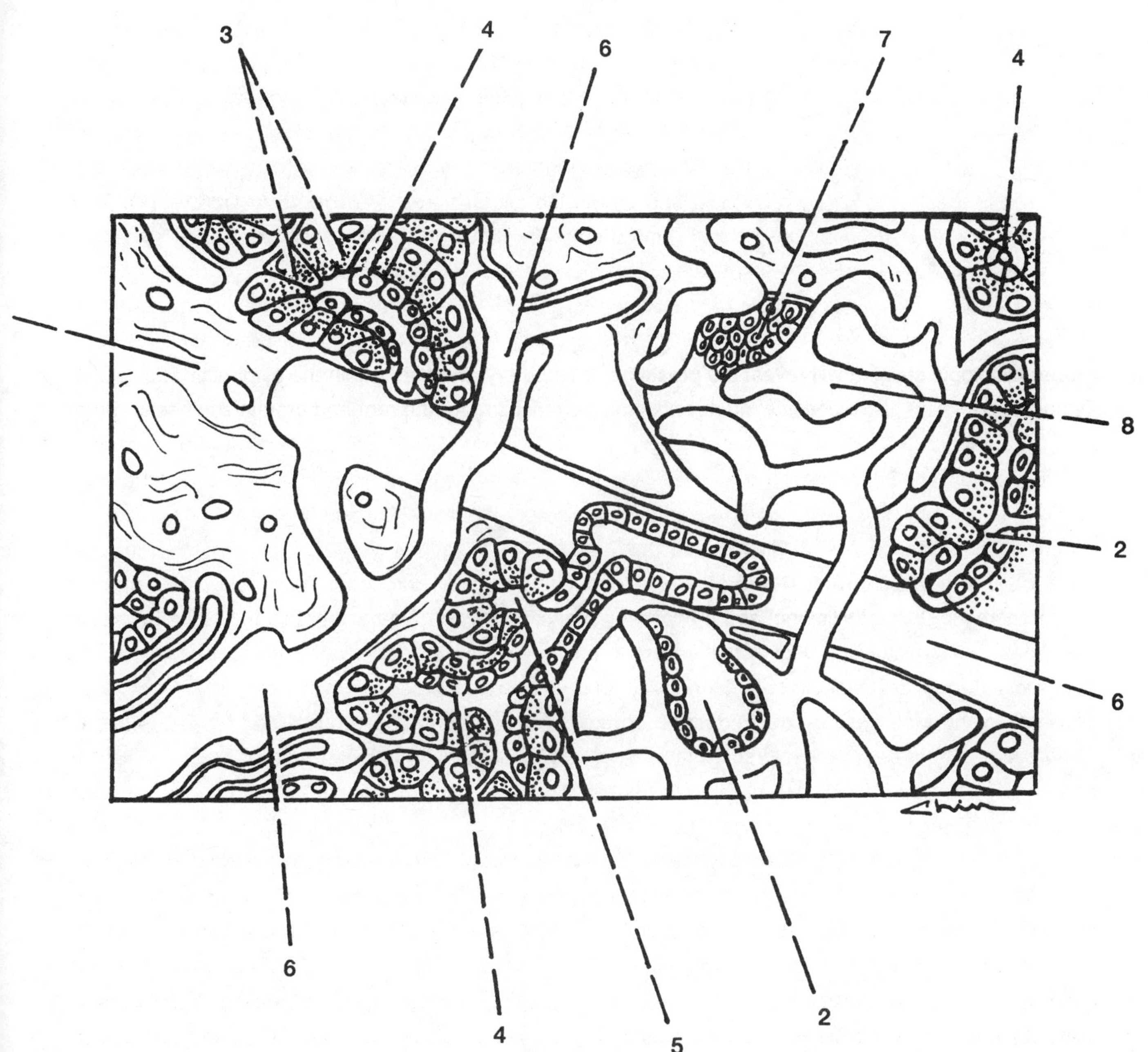
3
4
6
7
4
8
2
6
6
4
5
2